MW00463417

DeWitt Clinton and Amos Eaton

DeWitt Clinton and Amos Eaton

Geology and Power in Early New York

DAVID I. SPANAGEL

Johns Hopkins University Press

Baltimore

Johns Hopkins University Press
2715 North Charles Street
Baltimore, Maryland 21218-4363
www.press.jhu.edu

Spanagel, David I.
DeWitt Clinton and Amos Eaton : geology and power in early
New York / David I. Spanagel.
pages cm
Includes bibliographical references and index.
ISBN 978-1-4214-1104-0 (hardcover : acid-free paper) — ISBN 978-
1-4214-1105-7 (electronic) — ISBN 1-4214-1104-0 (hardcover : acid-
free paper) — ISBN 1-4214-1105-9 (electronic)   1. Eaton, Amos,
1776–1842.   2. Clinton, DeWitt, 1769–1828.   3. Geology—New
York (State)—History—19th century.   4. Geology—United States—
History—19th century.   5. Geology—Political aspects—New York
(State)—History—19th century.   6. Scientists—New York (State)—
Biography.   7. Politicians—New York (State)—Biography.   8. New
York (State)—Politics and government—19th century.   9. New York
(State)—Intellectual life—19th century.   10. Rensselaer Polytechnic
Institute—History.   I. Title.
Q143.E3S65 2014
557.47092'2—dc23
[B]   2013025010

A catalog record for this book is available from the British Library.

*Special discounts are available for bulk purchases of this book. For more
information, please contact Special Sales at 410-516-6936 or
specialsales@press.jhu.edu.*

Johns Hopkins University Press uses environmentally friendly book
materials, including recycled text paper that is composed of at least
30 percent post-consumer waste, whenever possible.

To the memory of Alan Lee Hankin (1948–2005)

Contents

Figures and Tables

Preamble

Before we begin to walk through the pages of the past, let us take a moment to determine where we are going and by what routes we might hope to arrive at our destination. The historical characters whose lives and work are featured in this study were generally curious, intrepid, and persistent people who did not seek merely to get by (or even get ahead) in matters of daily existence. These men (formal study of North American geology was at that time pursued almost exclusively by males) really hungered to make an imprint on the fabric of their world, both as public figures and as private seekers of scientific knowledge. The quest to plumb the secrets of Earth's history as a guide to material bounty motivated naturalists to endure intellectual frustrations and physical hardships. They rattled their bones traveling on rutted roads, wore through their boots scrambling over hills, streams, and rocky gullies, and tested their mettle investigating caverns and crevices made by natural or, when practical, by artificial means.

With great respect for the struggles and inventiveness of the early American republic's adventurers in New York politics and science, I ask for a considerable investment of patience and attention from the modern reader. To reach the level of sublime understanding, one must penetrate many layers of chalky sediment. Antiquated forms of speech and occasionally tendentious prose are part and parcel of this landscape, but the reward for following the paths of historical primary sources is well worth the trouble of immersion in the language, thoughts, and logic of that world. Walk carefully through the extended passages drawn from another time and place, contemplating the rationality and imaginative prowess of its inhabitants, and it becomes clearer how and why they found it useful or necessary to create changes we now take for granted, as well as introduce considerations that may now puzzle us. As we begin our journey, it helps to remember that during this time the United States lacked a unified national sensibility. Its citizens granted their allegiance to states first, and the possibilities of positive government resided most clearly in the encouragement of local conditions for individual prosperity.

Of the original thirteen states, New York was most blessed by physical geography and historical accident. Unlike any other political unit along the eastern seaboard (unless you include Lower Canada, the area we know as Quebec), New York State emerged from the War for Independence with easy access to navigable avenues into the interior of the continent and to well-established centers of oceangoing commerce. The watersheds of its major rivers were generally confined within the state's boundaries, and so it did not have to quibble with downstream neighbors about merchants' rights to carry the produce of New York farmers, miners, and manufacturers to the wider world. This single fortuitous circumstance provided the impetus for New Yorkers to pioneer many of the key tools that would ultimately prove essential to early republican nation-building endeavors: support by prominent men for its learned societies, the establishment of institutions of higher education, public financing of large-scale internal improvement projects, and the launching of efforts to complete comprehensive state geological mapping and natural history surveys to identify and assess valuable agricultural and mineral resources throughout the domain.

DeWitt Clinton and Amos Eaton

A Meeting Place for Waters and Students of Earth History

For a time, Albany, New York, was the crossroads of the world. Important people kept moving to, coming from, or passing through the neighborhood of the once sleepy state capital. Like many places in the early American republic, Albany's rate of population growth accelerated throughout the first three decades of the nineteenth century.[1] One key reason for this change was geography. Together with Schenectady and Troy, Albany was located near where the Mohawk River empties into the Hudson River. Despite their interior remoteness, these towns enjoyed convenient access to other desirable places. Eighteenth-century observers may have attributed the favorable natural situation of these formerly Dutch colonial settlements, which sat enviably at the head of a majestic interior valley, to the action of a benevolent divine Providence. Direct participation in global commerce, for one thing, was possible because measurable tidal changes reach up the Hudson River as far as Troy. Oceangoing vessels could therefore serve the needs of remote upstate farmers, and Albany's merchants could engage in trade upon the high seas.[2]

The leading citizens of New York's capital region were not complacent about their geographical advantages, and they were quick to explore and apply the levers of material change. Visionaries aggressively sought to augment nature's boon. During the first four decades of the nineteenth century, a celebrated series of revolutionary transportation innovations took place in the area, making Albany a prominent link in a rapidly multiplying network that carried not only valuable trade goods but people and ideas as well. For example, when astonished onlookers flocked to Albany's port in the summer of 1807, they saw the successful arrival and departure of Robert Fulton's *Clermont*. This "maiden voyage" inaugurated the era of steam-powered river travel in the United States. With the completion of a monumental 363-mile-long canal eighteen years later, New York State's prosperous commercial corridor expanded to include not just the placid Hudson but everywhere within navigable reach of the interior waters of the upper Great Lakes. A scant nine months after completing this major achievement, locals who were already spoiled by the convenience of the Erie Canal began to act on a hunger for something even better. They were frustrated with the slow pace of freight transit through the many locks that had been so laboriously built to circumvent the Mohawk River's rapid drop between Albany and Schenectady. In the spring of 1826, an application was granted to charter the Mohawk & Hudson Railroad, the first link in what would become the vast and powerful New York Central Railroad Company. Regardless of the railroad's promising future, the canal brought hundreds of new faces through on a daily basis, and the region changed with them. As one American historian puts it, Albany "in the grip of metropolitan industrialization" began to resemble Boston or Philadelphia, with a hundred new blocks of middle- and working-class housing sprouting up all around the old business district, while Troy was transformed from a small mill town into "the foremost iron town in the Republic."[3]

Along with material innovation came political clout. As the legislative seat of the Empire State, Albany enjoyed a proximity to power far beyond that of any other American town its size. By 1817, when ground was first broken on the project to excavate the Erie Canal, New York State had already supplied three of the nation's first six vice presidents. All three were members of Thomas Jefferson's Democrat-Republican Party: disgraced Aaron Burr, venerable George Clinton (deceased by 1817), and incumbent Daniel Tompkins. New York had also fielded its first presidential contender. After fighting erupted in 1812 between the United States and Great Britain, Clinton's nephew DeWitt Clinton mounted a bid to unseat James Madison. The popularity of Jefferson's Virginian successor diminished rapidly among northerners who saw little to gain and much to lose in the war. Garnering the support of New England's Federalists as well as fellow Republicans in most of the northern states, DeWitt Clinton fell just short of a stunning victory over the wartime incumbent that fall.

Though his presidential aspirations were frustrated, DeWitt Clinton remained a powerful figure by serving as New York City's mayor. He used the remaining years of the war to reconsolidate his spent political capital, in part by capably defending the city and in part by expanding his reputation as an intellectual, cultural, and scientific leader. Many of New York City's most illustrious civic institutions date from this era, and DeWitt Clinton had a prominent hand in their establishment. When Virginia's James Monroe selected New York's sitting governor to be his running mate in the uncontested election of 1816, Clinton was poised to harvest the opportunity created by Tompkins's departure from Albany. Winning the special election by an overwhelming margin in the spring of 1817, DeWitt Clinton relocated his power base to Albany and proceeded to make the construction of the Erie Canal the centerpiece of his political career.[4]

Most historians of the early American republic would explain the birth of the Erie Canal in terms of the combination of geographical good fortune, civic enterprise, and adroitly placed political power that I have outlined. However, a few historians of American science and technology have probed more deeply to examine the circumstances within which Albany, Schenectady, and Troy played host to a substantial local flowering of American scientific talent and accomplishment. Albany native Joseph Henry, for example, became the young nation's first great experimental physicist (unless you count Benjamin Franklin). Henry began his technical education at the Albany Academy in 1813, just six years after this preparatory school was founded. As an adult, Henry would gain worldwide renown as a codiscoverer of electromagnetic induction along with Englishman Michael Faraday, and he would go on to promote American science as the first secretary of the Smithsonian Institution.

An institutional infrastructure for the natural sciences scarcely existed anywhere in the early American republic at the time of Henry's birth in 1797. That year, however, Albany's unlikely emergence as a center of intellectual leadership began when the Society for the Promotion of Agriculture, Arts and Manufactures left New York City, where it had been founded in 1791, and moved to the state's capital. Here it was reconstituted as the Society for the Promotion of Useful Arts (SPUA). With the establishment of a New York State Board of Agriculture in 1819, the SPUA diverted its focus from the "useful arts" and made a deeper commitment to the study of technical fields within the earth sciences: mineralogy, botany, and geology—or, as contemporaries would have called these endeavors collectively, the practice of natural history.

In 1823, the Albany Lyceum of Natural History was launched with a mission to bring the study of these kinds of materials into greater public awareness. Natural history embraced some of the most exciting and profound questions humans could ask in the post-Enlightenment era. Scientists of the early nineteenth century were fascinated by questions such as: How old is the Earth? When did time begin? How might the passage of time have shaped and reshaped the landscape since its original creation? How were visually dramatic features (like great chasms, mountains, and waterfalls) or particularly valuable ones (like beds of salt, coal, and gypsum) formed? Was the world designed by an intelligence who had our needs in mind? Did the native species of rocks, animals, and plants that Europeans found in North America arise independently, or do all global natural phenomena share common origins and histories?

Answers to these large questions of natural history might be sought anywhere in the world, but in the newly formed United States, New York was an especially favorable place for such scientific inquiry. By the mid-1820s the necessary infrastructure and places of education had sprung up, just in time to inspire the young minds of Joseph Henry's generation and those that would follow. In 1824, the Albany Lyceum merged with the SPUA's already impressive collection of high-quality specimens to form a new museum called the Albany Institute. Meanwhile, nearby institutions of higher education began to offer instruction in the natural sciences and related technical fields of engineering, subjects that until then could not easily be studied outside the intellectual capitals of Europe. One prominent example was the U.S. Military Academy founded by the federal government in 1802 on the heights overlooking the Hudson River halfway between Albany and New York City. West Point was intended both to train officers and to provide the army with engineers. Within a quarter century, Troy's Rensselaer School would join West Point, becoming the first institution in the United States where civilians could obtain an education in engineering.

Various kinds of scientific training were already available in New York even before the turn of the nineteenth century. For example, rudimentary instruction in the theory and practice of botany, chemistry, geology, and mineralogy was available at New York City's Columbia College. Many of New York's first generation of political leaders attended this traditional, classics-heavy institution, and the modest exposure to and respect for learning in the sciences they gained there would have a profound social and cultural influence on the early American republic. Beginning with the hiring of Benjamin Silliman in 1806 to teach at Connecticut's Yale College, opportunities to study branches of natural history proliferated across New England. Yale-trained students were hired over the next decade to introduce isolated scientific courses at Vermont's Middlebury College, Maine's Bowdoin College, and Massachusetts' Williams College. In 1828, Schenectady's Union College (founded in 1795) instituted a scientific curriculum designed to rival Silliman's program at Yale in its comprehensiveness. A resident from Albany could obtain a natural science education at Union or a practical technical education at Rensselaer, an undertaking markedly different from the classical literary training that was the main focus of most other American colleges.

The Albany region now competed directly with other established centers of

American scientific expertise: Boston, New Haven, New York City, and Philadelphia. The culmination of early public investment in the study of natural history
came in 1836 with the establishment of the New York Natural History survey, an
ambitious project to map and take an accurate inventory of natural resources located throughout the entire state. This survey provided the first viable institutional
home for professional scientific research. There, a handful of scientists could practice botany, geology, and paleontology. It is no accident that when the first group
of America's professional scientists gathered to organize themselves in 1840, they
met in Albany to form the AAG (American Association of Geologists), later to be
renamed the AAAS (Association for the Advancement of American Science).[5]

WHO WAS AMOS EATON?

Amos Eaton (1776–1842) was once considered a major figure in early American
science, and his life experiences provide a useful lens for examining the effect that
geological knowledge and practices had on New York State's politics, prosperity, and
culture prior to the 1840s. Unlike the ingenious and successful Joseph Henry, Eaton's life and accomplishments in science, though highly original, ambitious, and
innovative, were often frustrated by circumstances and occasionally ludicrously
incorrect. Despite his many tribulations, Eaton played an essential role in the
story of how the science of geology became intertwined with New York's rise to
power and regional intellectual significance.

Eaton's career was buffeted by extremes of failure and success. Initially educated in the classics at Williams College and channeled toward a life practicing
law, Eaton's love for studying botany in his spare time proved to be the only bright
spot in his first decade of adulthood. Framed by political enemies and wrongfully
imprisoned for forgery, Eaton was ripped from his young family and the idyllic
small-town life that he had carved out in Catskill, New York, in 1811, cast into
Newgate prison in New York City's Greenwich village, and sentenced to serve a
term of life at hard labor. Had that verdict been the end of this story, only the
most specialized historians of early American botany would recognize the name
of this obscure self-educated field botanist who self-published a modest manual
for the study of New York plants in 1810. However, through a combination of ac-
cidental encounters, assiduous study, and the miraculous good fortune of a happy
outcome to the War of 1812, Eaton's freedom was restored in 1815 under a general
amnesty. The thirty-nine-year-old ex-convict set out immediately to make a fresh
start based on his love for natural history. He would start not only by practicing
botany but by seeking additional training to become an accomplished mineralogist and geologist.

Harboring lofty and expansive professional aspirations in science, Eaton acquired the skills he would later need when opportunity knocked. With the financial and intellectual support of an aristocratic patron,[6] Eaton would conduct and publish internationally respected research on the mineralogy and geology along and beyond the Erie Canal, precisely when its initial bed was being excavated and constructed across New York State. These achievements enabled Eaton to assemble a comprehensive detailed geological map, the first of its kind to be made for such an extensive region.[7] In 1824, Eaton persuaded Stephen Van Rensselaer to use a bit of his vast wealth to establish the Rensselaer School, a radically new institutional form of higher education. Conceived as a college of *practical* science and engineering, the school embodied Van Rensselaer's vision of preparing and placing scientifically-trained men throughout the far reaches of New York State.

During his fifteen-year tenure as the first senior professor at Rensselaer, Eaton would develop and perfect his highly interactive style of pedagogy and encourage all of his students not only to master received ideas but to boldly set forth to create new scientific facts, theories, and methodologies based on their own observations and deductions. Eaton's unorthodox teaching style of forcing his students to actively engage in lecturing to each other fit perfectly with the school's mission. An 1828 Rensselaer Trustees' report to the New York State Legislature laid out and justified Eaton's pedagogical approach as follows: "in the place of the professor [the pupil] necessarily acquires a knowledge of the principles of the science on which he lectures; while the experimental demonstrations of the laboratory render him familiar with the practical applications of those principles to agricultural and manufacturing operations."[8]

The technique proved to be extremely fruitful and effective, especially for inspiring disciples capable of exceeding their teacher in their chosen scientific fields. By the end of his life, Amos Eaton could claim credit for having directly nurtured the scientific curiosity of an astounding number of important nineteenth-century Americans, people whose gifts and training often carried them far beyond the boundaries of New York State. Among Eaton's famous protégés were amateur botanist and acclaimed writer William Cullen Bryant; professional botanist John Torrey; pioneering female educators Almira Lincoln, Mary Lyon, and Emma Willard; geologists Ebenezer Emmons, Edward Hitchcock, and Douglass Houghton; paleontologist James Hall; and physicist Joseph Henry.

Despite Eaton's significance to the history of American geology, he has not received much positive attention from modern scholars in the field. For example, historian of Jacksonian science George Daniels, in his influential 1968 book, clumped Eaton, Silliman, and Hitchcock as a trio of essentially derivative early

American geologists who spent their careers as disciples of Abraham Gottlob Werner, an eighteenth-century German mineralogist who promoted a systematic understanding of rocks. According to Wernerians, Earth's rocks successively precipitated from an original universal terrestrial ocean, according to a logical sequence of hardness and chemical solubility, from the primitive rocks like granite through the secondary sedimentary rocks like sandstone. In the early nineteenth century, these views were widely taught and referred to as the Neptunist theory. According to Daniels, Hitchcock and Silliman evolved somewhat further in their views than Eaton. They accepted an alternative doctrine, late-eighteenth-century Scottish agriculturalist James Hutton's "theory of central fire," also known as Vulcanism. Only Eaton's use of mineral contents to date strata seemed to Daniels to be an implicit departure from Wernerian dogma.[9]

Daniels ultimately dismissed Eaton's career and achievements because they predated the time when American science became recognizably professional. By using the term *professional*, Daniels specifically contrasted the methodologies Eaton inherited from those promulgated by the subsequent generation of earth scientists: "Physical classification [of specimens] since the time of Werner had been arrived at on the basis of lustre, hardness, color, brittleness, and other readily ascertained characteristics. The new system, on the other hand, required a knowledge of chemical analysis. In geology, the crucial event was the increasing use of fossils as guides to dating strata—again, a task requiring specialized knowledge and wide agreement upon principles of nomenclature."[10] Daniels somehow neglected to note Eaton's pivotal role in this process of professionalization. As a matter of fact, when the French geologist Charles Lesueur visited Albany in 1820, he instructed Eaton on the use of characteristic fossils to identify strata, "thus laying the foundations for future work of the rising school of New York paleontologists."[11]

Some more recent works on the history of American geology, such as those by Michele Aldrich and Paul Lucier, present stronger historical arguments about the challenges Eaton overcame and about his legacy.[12] Their discussions of his theoretical and practical influence on the field demonstrate a higher regard for the enormous logistical and intellectual obstacles that New York State's geology presented to anyone attempting comprehensive or systematic analysis of its contents and configurations. Nevertheless, because Aldrich and Lucier focus their respective studies on work done by the generation largely composed of Eaton's students and junior colleagues, they still tend to characterize Eaton himself as a relatively primitive precursor to the era of professional American geology. The originality and fruitfulness of Eaton's own achievements as a thinker and practitioner of mineralogical and geological classification schemes remain largely unexamined and implicitly undervalued, as do his innovative uses of stratigraphy and extrapolation in territorial mapping, and his ingenious and painstaking efforts to develop a mechanistic understanding of the dynamics of large-scale flooding in order to account for the variety of debris deposition (which would later be understood as consequences of cycles of continental glaciation). Respectable for its historical accuracy but equally devoid of attention to Eaton's theoretical prowess, Andrew Lewis's recently published study emphasizes only the utilitarian promises that Eaton frequently made in order to court public support for scientific investigation.[13] Least useful are other scholarly works about Eaton that show little regard for historical facts.[14]

Beyond Eaton's insufficient treatment in works focusing on the history of science, there are few indications that modern American historians generally appreciate the contemporary relevance geological knowledge had for the broader social fabric of the early American republic. Occasionally, art historians speculate that landscape painters like Thomas Cole must have been acquainted with scientists such as Silliman and Eaton, but it is exceedingly rare to find any explicit mention of Eaton at all in works not directly concerned with the history of American botany or geology.[15] Otherwise excellent books from recent years covering the history of the Erie Canal, the environmental history of early New York State, and even the history of natural history in the United States (all discussions in which Amos Eaton is worthy of some mention) continue this general pattern of neglect.[16]

Eaton's influence as an educator and role model for American science was simultaneously monumental and circumscribed. Despite his extraordinary career as a popular lecturer and teacher, the shelf life of his original scientific theoretical output was limited. While his ideas about the dynamic role of flooding throughout Earth's history were avidly read and discussed in their own time, and his attempts to systematize botanical and geological nomenclatures were generally greeted with respect and enjoyed wide dissemination, they were subsequently discarded, even by his most devoted students. Eaton had trained them to reject whatever seemed antiquated or dubious, and they applied this principle to their mentor's teachings.

Admittedly, a comprehensive scholarly biography of Eaton is long overdue; the only substantial works include Harlan Ballard's 1897 memoir for the Berkshire Historical and Scientific Society and Ethel McAllister's excellent but not very widely circulated 1941 University of Pennsylvania education dissertation.[17] A balanced reassessment would analyze Eaton's professional ambitions for American natural history, where he made serious contributions to the development of

botany, geology, and mineralogy. Rather than pursue this appealing but monumental biographical task, however, I have chosen to feature the geological ideas
and practices that New Yorkers imagined and discussed during Eaton's lifetime.
In particular, I emphasize the theoretical and philosophical implications of contemporary attitudes toward various hypothetical agencies of geological change
throughout Earth's history, such as God, water, fire, time, and ice. Where he
plays an active role in these discussions, I follow Eaton rather closely through
his writings, experiences, and relationships. Elsewhere in this book, his life story
drops away to permit a broader survey and analysis of those surrounding political,
economic, technical, and cultural circumstances that enabled geological investigations and discourse to inform and reshape the society of New York State.

AMOS EATON AND THE GEOLOGICAL IMPERATIVE

The New York system of geological stratigraphic nomenclature, which was conceived and elaborated by Amos Eaton and his students, provides an excellent metaphor for the larger set of historical processes this book examines. Geology was
the science that enabled New Yorkers to cultivate power through the knowledge
of place. The following chapters examine how knowledge about geology acquired
this imperative quality in the New York context. They also show how, despite
its many impressive achievements, this localized knowledge system ultimately
fell short of its promise. Eaton's New York system, though widely admired and
imitated, did not in the end provide a globally applicable model for transforming
knowledge into power. But the practical appeal of his hands-on approach to geology continued to flourish, even when his pet theories and nomenclature preferences were superseded. The relevance of geology to nation building was easy for
contemporaries to see. Empires of facts, peoples, and places always depend on
local circumstances, and they ignore particularities at their own risk. The practices of natural history operated in concert with individual aspirations for fame
and state interests in the cultivation of knowledge. Knowledge about the shape,
structure, and contents of the land came to be perceived as a critical ingredient
in successful settlement and a promising route to the harvest of what would otherwise be hidden stores of natural wealth.[18] In other words, because of New York's
example, science became so widely understood and utilized (as being a key to
effective natural resource exploitation) that most American historians have failed
to question how or explain why this social investment in scientific authority and
technical practice occurred when and where it did.

On a more lofty intellectual plane, theoretical achievements by home-grown
natural philosophers also helped validate cultural self-confidence. But how might

efforts to advance American reputations and achievements in the earth sciences matter to larger questions of early American history? Writers on the subject of national identity have focused on the processes by which people come to believe that they have something essential in common, and how they acquire and utilize various means of communicating this shared vision to the wider public.[19] If the process of national identity formation largely involves strangers who collectively discover a sociocultural point of commonality, based primarily on a shared language and a nationalistic awareness cultivated by an emerging media, how might "nature" or "place" matter? Some historians emphasize the role that land (or "nature," broadly construed to include both geography and natural resources) has played in abetting forces of nationalism, especially in the grand scheme of European acquisition of global wealth and power.[20] But others suggest that the differences in modes of scientific knowledge about nature matter more.[21] Regardless of these distinctions, American natural historians operated not only in a scientific intellectual context but also under the pressures of intense political and practical economic considerations.

The starting point for everything in North America was the land itself. Americans flung themselves into the quest to occupy and own their own private portions of land. Power and wealth derived from the ability to possess land and exploit its resources. Geographical circumstances had contributed far more profoundly than manpower, weaponry, or military strategy to success in the war for independence from Great Britain. Subsequently, gaining knowledge about the contents and the structures of nature automatically became an essential precursor to imperial conquest. Natural historians provided important services by developing a regime of scientific authority in alignment with political power. I have chosen to focus this study of geological knowledge and its political, material, and cultural effect on New York because the idea of harnessing natural history to the task of empire building was formulated, tested, and promoted through this particular state's example. This is not to say that the geological imperative was always the result of conscious or deliberate efforts. Scientific practitioners were sometimes engaged in necessary but mundane tasks like locating mineral resources and mapping topography. At other times they were absorbed in dreamy attempts to invent systematic theories of the Earth or to create schemes for naming its various constituent creatures and parts. This book examines how scientific activity contributed to the achievements of an impressive array of prominent figures in New York politics and culture. Through this narrative, the early American republic can be better understood as a social experiment whose cultivation and deployment of natural knowledge depended essentially upon the practice of geology.

Patrons and practitioners of the earth sciences were concerned about more than just the technical soundness and practical applicability of local geological knowledge; they hoped to garner international respect. After Alexis de Tocqueville visited the United States during the 1830s, he published an influential study entitled *Democracy in America*. Tocqueville regarded science as an indicator of general cultural advancement, and his assessment of Jacksonian America in this respect was less than flattering: "It must be admitted that few of the civilized nations of our time have made less progress than the United States in the higher sciences or had so few great artists, distinguished poets, or celebrated writers."[22] Though he acknowledged the cleverness with which science had been applied to serve practical needs, Tocqueville ascribed to America a pervasive "lack of genius," which he justified by citing its paucity of theoretical scientific achievements. Anxieties about cultural and intellectual inferiority certainly did trouble Americans. Scientific thinkers writing about natural history were faced by challenges on all sides. Even well-to-do gentlemen like Clinton and Van Rensselaer who were curious about nature often stressed the practical value of knowledge about native rocks, plants, and animals when they sought to inform the public about science. A peculiar rhetoric thus emerged in the earth sciences, emphasizing tangible assets, such as mineral and agricultural resources, above theoretical breakthroughs. At the same time, however, these men sought recognition for their discoveries and the respect of their foreign counterparts. Despite the rhetoric, American natural historians also cared intensely about matters of theory and methodology; they were almost never mere prospectors, sifting for "facts."

This book argues that natural history played a critically valuable role in the intellectual and political life of New York State. Private, and later public, support for science was linked to expectations of prosperity as a result of relationships that grew up among these different kinds of practitioners and consumers of geological knowledge. Chapters 1, 3, 5, and 6 focus primarily on technical ideas and activities, while chapters 2, 4, 7, 8, and 9 branch out to consider how these ideas and activities stimulated or constrained contemporary changes in political, economic, and cultural conditions. Sliced another way, the initial third of the book sets the stage by introducing geographical and personal dimensions of historical contingency. It begins by focusing on the parallel careers of early New York political leaders DeWitt Clinton and Stephen Van Rensselaer and only then delves directly into that of Amos Eaton (whose life in science required the support and sponsorship of these two key patrons). Eaton's peculiar career invites the historian of early American science to pursue multiple threads of analysis. He was a tireless practitioner of field research, as well as a well-traveled popular lecturer, innovative educator, and influential mentor for the next generation of professional American botanists, geologists, and mineralogists. Finally (and this is the dimension that few recent scholars have examined closely), Eaton was a provocative geological theorist and system builder. An irrepressible teacher, writer, and self-publicist, he devoted the second half of his life to resolving global questions of Earth's history and improving the technical practices of geological science, including innovations in systematic nomenclature and in the production of visually instructive geological profiles and maps.

The middle third of the book explores in detail how New York's examples of privately and publicly sponsored scientific research and teaching fulfilled the practical demands of knowledge and skill that were needed to build a regional infrastructure for transportation and commerce. For internal improvements programs to take off, governments needed to see a return on investment of public monies. New York's stunning success with the Erie Canal persuaded many other states to follow suit. Replicating that success, however, depended entirely on a particularly fruitful and synergistic relationship between geology, hydraulic engineering, and demographic mobility. In the absence of this delicate balance, the harsh dictates of topography, politics, and economics almost always spelled disappointment or even failure for the ambitious canal projects launched in other parts of the country during the ensuing decades.

The final third of the book examines larger linkages between the knowledge produced by New York's geologists and the content and thematic attitudes expressed in contemporary prose, poetry, and fiction, landscape and history paintings, and religious movements in antebellum American society. To what degree did New York's leading figures in these "unrelated" fields of creative, intellectual, and spiritual endeavor have direct interactions with geologists like Amos Eaton? What constraints or historical circumstances imposed limits on geology's influence and authority in the broader culture? How did the social authority of science negotiate to gain wider public understanding and respect in the early American republic? These chapters survey the social and cultural consequences of attempts by Eaton and his contemporaries to uncover and interpret New York's geological history.

Though their careers as researchers and theoreticians in the first decades of the nineteenth century were largely eclipsed by the achievements of the next generation, the naturalists employed and sustained by Clinton and Van Rensselaer nevertheless launched an ambitious program to comprehend the natural history of New York, and thereby set a strong example for those who would become the first generation of American professional scientists. The exploits of these early investigators of nature are all the more illuminating, given that these men were not yet so professionally disciplined as to avoid thinking across or speculating beyond what would later be recognized as distinct boundaries among the fields of natural history. An Enlightenment ethos generally encouraged inquiry and intellectual discourse, even across nascent disciplinary boundaries, throughout the early nineteenth century.

American science developed in a cultural setting far different from that experienced in Europe, and historians examine the normative implications of these differences. One fine article notes that Americans in the 1820s and 1830s saw scientific knowledge and practice through a popular democratic lens, rather than an exclusive elitist one. Barriers to scientific authority and the delayed development of disciplines in the American context can therefore be studied by looking at popular forms of science.[23] This book poses the larger and more concrete challenge of explaining how and why an "Empire State" and a professional scientific infrastructure were co-created in New York. It was no accident that geology was the first American scientific discipline to professionalize, to acquire substantial public funding, and to have an influence on American fine arts and moral philosophical discourse. But to understand the contingencies that invested all these relationships with meaning, it helps to see the world through early nineteenth-century eyes, armed only with what could be known and imagined about geological time, structures, and processes by people who had barely scratched the Earth's surface.

PART ONE

Exploring New York State

CHAPTER ONE

Invitations to Study
the Earth's Past

Scenery, indescribably grand. Water on the Mohawk looked "dark
and troubled." Huge rocks on either side of the road [leaving Little
Falls], in various places, could perceive big holes evidently worn by
the friction of water. Look at Mitchell [*sic*] again. He will tell you,
how big the Mohawk once was; where, when and how an immense
lake, which covered all the country, now known as the German
Flats, "Thaw'd and resolv'd itself into a dew." When the Hudson
broke through its rockbounded barrier in "the Highlands," and
formed its present course to the Ocean.

—*A Knickerbocker tour of New York State, 1822*

AN OVERVIEW OF NEW YORK GEOGRAPHY

New York's physical geography attests to the marvelous history of the planet.
Along its eastern boundary can be seen the grandeur of an impossibly ancient
tectonic-plate collision, which upthrust the now bucolic Taconic Mountains so
violently that they once stood as tall as today's Himalayas. On the other side of the
state, the power of so mundane a force as ordinary fluvial erosion is displayed in
the mighty flow of the Niagara River as it drops down a small step to thunderous
effect. Everywhere in between lies evidence of many other historical geological
processes. Broad tracts containing fossil-bearing sedimentary rocks attest to extensive durations of marine incursions upon land now far from the oceans. An otherwise completely confusing jumble of soils, sand, and rocks of all sizes and types
date from more recent periods of extreme climate change, in which the advances,
retreats, and melting of continent-covering sheets of ice not only shuffled surface
materials but also created many of the state's most prominent lakes and valleys.

Much of what is now known about the Earth's history has been discovered
and refined only within the past 170 years. In 1800, the modern ideas about cli-

mate dynamics or structures of the Earth's interior that now provide fundamental bases for widely accepted accounts of our planet's geological history had not yet been developed. At that time, nobody would have imagined trying to analyze the dynamics of continental glaciation. Nevertheless, early nineteenth-century American geologists were fascinated by basic problems of decoding surface clues that might indicate anything about the hidden structures beneath to help identify the whereabouts of potentially valuable minerals. They felt compelled to explore remote terrains and to weave imaginative rationales from their meager harvests of hard-won data. Later generations have tended to denigrate the quality of those first provisional explanations, which were really impressively resourceful when one considers that early scientists virtually had to invent the tools of investigation simultaneously with the theories that could guide their exploration.

The range of early American scientific interest in nature extended boundlessly across the kingdoms of animal, plant, and mineral, but the natural historian was constrained by two factors. First, the geological record lay mostly hidden from view and was therefore inaccessible to all but the most adventurous and energetic travelers. Second, a miserly handful of speculative theoretical frameworks were all that European thinkers had developed and promoted during the preceding century, and these were prejudiced in important ways by a dogmatic Judeo-Christian scriptural tradition. Even so, the American geologist had a few things in his favor: persistence, a willingness to blaze new trails, a new land not yet known to science, audacity, and a healthy disregard for European authority. These advantages combined especially well for those who studied natural history in New York.

Early observers of New York's topography reported their encounters with a wildly diverse variety of landforms and natural features. The northern part of the state contained the most rugged mountain province in the northeastern United States, the Adirondack Mountain dome. A ring of lowland areas circumscribed this persistently impenetrable wilderness, however, with natural navigable waterways roughly delineating all four sides of the bounding quadrilateral: Lake Champlain to the east, the Mohawk River valley to the south, Lake Ontario to the west, and the Saint Lawrence River valley to the north. Most of the rest of the state consisted of a gently rolling topography of hills, valleys, and waterways. The curiously eroded Allegheny plateau dominated the western and central regions, with the intermittently interrupting ridges of the Catskill and Helderberg Mountains forming an eastern frontier. The north-south Hudson River valley cut its dramatic gorge through the Hudson Highlands toward the sea, and the still-forested remnants of an ancient (and formerly stupendous) Taconic Mountain chain traced the eastern boundary of the state. Finally, Long Island rode offshore as an isolated specimen of the Atlantic coastal plain.

Mysteriously, New York's physical geography appeared to result from dramatic and forceful incursions of water upon the land. As historians of European geology have observed, the careful study of physical geographic features was, by the late eighteenth century, likely to arouse in natural history investigators an awareness of "deep time." Writers often cite the time-expanding effect of theoretical works such as *Epochs of Nature*, published in 1778 by the Frenchman Georges-Louis LeClerc, Comte de Buffon, and the eternal Earth machine that was first framed in modern geological terms by Scotsman James Hutton in his 1795 *Theory of the Earth*.[1] But curious early nineteenth-century visitors to France's prolific fossil basins or even Scotland's billion-year-old igneous rocks could not have experienced so visceral an impression of our planet's antiquity as did those who were afforded the opportunity to view the natural wonders of New York State.

New York's prominent geological features would naturally provoke thoughts about Earth's history. Most of the state is covered by sedimentary rock formations that date from hundreds of millions of years ago, when the Canadian Shield terrane formed the nucleus of a Precambrian North American continental mass. All the rocks to the east and south of Lake Ontario (except the Adirondack Mountain province) were originally formed offshore as part of the continental shelf. But these rocks have not been left undisturbed throughout the intervening eons. For example, during periods of glacial advance, the gently tilted plateau that forms the western part of the state was creased by a series of gouges. When the ice melted, these depressions filled to become freshwater lakes, comparable to the North Sea and the Baltic Sea in their extreme depths. These narrow Finger Lakes are oriented primarily from north to south, and though typically only a mile or two across at their widest points, two of these lakes approach forty miles in length. To the west of the Finger Lakes lie two of the only significant rivers in the contiguous United States that flow from south to north. Both the Genesee and the Niagara cut gorges through layers of sedimentary rock, exposing broad vistas of North America's geological past. These views would only be surpassed late in the nineteenth century with the discovery of the Grand Canyon of the Colorado River. Finally, Lakes Erie and Ontario presented awe-inspiring expanses of freshwater that seemed oceanlike to European American settlers accustomed to living among forested hills and valleys with limited horizons.

Because New York hosts all these peculiar bodies of water, spectacularly eroded landforms, and dramatic bedrock exposures displaying nearly horizontal layers of rocks replete with marine and freshwater fossilized organisms, the earliest natural history observers hypothesized the former existence of a vast inland sea whose size dwarfed all of the Great Lakes combined. From the 1840s on, professional geologists attributed many of these remarkable features to complicated movements by the massive glaciers that covered major portions of New York State intermittently from about two million years ago until as recently as twelve thousand years ago. For example, the plain that crosses the western part of the state rises gently to an apex, forming a perfect natural highway parallel to and about ten miles south of Lake Ontario's shoreline. The Ridge Road, as it is called virtually throughout its length, traces the former shoreline of what modern geologists have dubbed glacial Lake Iroquois, a body of water that existed when all of Lake Ontario's water was blocked by ice at the end of the last episode of continental glaciation, prior to when the Saint Lawrence escape route opened up. Postglacial melt in its various configurations explains "hanging" beaches, wave-cut cliffs, interior dead-end river deltas, sand dune forms, and other lakeshore features now observable far from the shores of Lake Champlain (glacial Lake Vermont), in various spots throughout the upper Hudson and lower Mohawk River basins (glacial Lake Albany), and across the lower Connecticut River valley (glacial Lake Hitchcock).

This fruitful glacial explanation, however, was virtually unknown to the world prior to July 1837, when the famous Swiss naturalist Louis Agassiz argued that continent-sized ice sheets must have operated in the past. He embraced this novel idea in order to account for an otherwise puzzling feature that had been widely reported throughout Europe and North America. Wherever extremely hard bedrock surfaces were exposed to view, peculiar sets of parallel scratch marks could be seen. In the Catskill Mountains and along the tops of cliffs in the Genesee gorge, these scratches tended to run from north-northwest to south-southeast. Agassiz noted how active glaciers in his native Alps leave similar traces of their passage, and this analogy provided the key to his revolutionary ice age (*Eiszeit*) theory.[2] Prior to this suggestion, catastrophic outbursts of flooding (diluvialism), heat (volcanism), and lateral pressure (earthquakes) were the only known causes geologists could point to as capable of dramatically reshaping terrain. Ice had only recently been recognized as a significant geological agent in part because it operated so comparatively slowly.

Living glaciers were known to transport rocks and soil. This fact made the terrain of New York State—as well as New England to the east and parts of Ohio, Indiana, Illinois, and Wisconsin to the west—textbook examples for the study of postglacial geology. The entire landscape of the Great Lakes region was bestrewn with extraordinary boulders. Far more noticeable than the parallel scratches, which one had to seek out to recognize, these boulders, also called *rocking stones*, fascinated the public, particularly in deforested places where Europeans had used up the wood for heating fuel and building materials. Settlers called these rocking stones *erratics* because of their apparent disregard for either the circumstances of local geography or local bedrock geology. Some were precariously perched atop rocky promontories, while others were found mysteriously half-buried in hillsides and valleys. Erratic boulders can weigh up to several hundred tons and invite deep contemplation of the forces that shaped the landscape. Ever since humans began to inhabit the northeastern United States, erratic boulders have attracted interest and provoked awe. They are just too startling and impressive to ignore.

Amid the hills and valleys where one could observe these imposing boulders, there also remained scattered piles of what is now recognized as glacial "drift." Rocky debris and soil scoured from elsewhere had over time been dumped by the leading edges of the ice sheets, to form snakelike mounds that glaciologists would later call *terminal moraines*. In other places, the clumps of clay, sand, and gravel were dropped by melting ice, only to be overrun again by resurgent overriding ice, forming small oblong mounds called *drumlins* (the name derives from the Irish word for "little hill") that look almost artificial in their oval symmetric perfection. These drumlins are sometimes isolated, but they can also occur in groups giving a "basket of eggs" appearance to the local topography. On the one hand, massive erratic boulders were a curiosity for travelers but mostly a nuisance to farmers. On the other hand, the seemingly random caches of clay and sand found among the moraines were potential treasure troves. Wherever found and quarried, these valuable deposits sustained highly profitable brick-making and glass industries.

Elaborations of the ice age theory would eventually help to explain why the Mohawk River valley is so broad and level. According to modern geological analyses, the melting ice covering the Great Lakes Basin would at first drain southward across the Midwest via the Ohio River valley to the Mississippi River during the interludes between glacial maxima. But whenever the ice dam blocking the Mohawk River valley was sufficiently melted or partially displaced by accumulated water pressure, central and eastern New York State would suddenly receive the entire outflow of the Great Lakes watershed. In this manner, geologists now speculate, rapidly moving water blasted a direct pathway to the Atlantic Ocean through the Mohawk and Hudson River valleys until re-encroaching ice again diverted the flow of water south and west or until the continued recession of the continental glacier further northward finally reopened the broader outlet to the sea through the Saint Lawrence River valley.

The advances, rather than the retreats, of those continental glaciers were responsible for that other remarkable pattern of erosion across New York State, the Finger Lakes. Separated by elevated strips of intervening land, these lakes do not bear the U-shape profile of a typical Alpine glacial valley but instead present a spectacular convexity wherein the valley sides actually steepen as they approach the shoreline. To explain this fjordlike appearance, modern geologists hypothesize that this streamlining of the upper slopes was accomplished by ice sheets overriding much thicker tongues of glacial ice. Early twentieth-century geologists established that the Hudson River gorge through the Highlands near West Point was produced in much the same way, whereby a fjord was ripped through layers of bedrock nearly one thousand feet deep. Subsequent deposition by the river has since refilled this glacial gouge, so that a bed of sediment over nine hundred feet deep now resides beneath the river's bottom.

Students of New York geology had observed virtually all of these fantastic features as early as the turn of the nineteenth century, but they had recourse to almost none of these explanations. Nevertheless, it was clear enough that the Earth's history was written upon the land right before their eyes. They ached to gain, improve, and even invent whatever tools of geological literacy and justifiable scientific speculation they needed to assemble an accurate account of that history. Though it may seem abstract, almost ethereal, to worry about how old the Earth is and to ponder what different conditions a location may have once experienced, for these early nineteenth-century Americans, geology was among the most "applicable" of the natural sciences. They understood that if the Earth has a history, there may be discernable reasons why certain valuable substances occur in certain places and not in others. Knowing what to seek and what not to seek, where to dig and where not to dig, would make all the difference between a harvest of riches and a fruitless investment of precious labor.

NIAGARA FALLS: EARTH'S CHRONOMETER?

Long before anyone worried about North American geology's implications for prevailing European belief systems, interested writers began publishing reports of superficial geological observations and speculations. Curious travelers visited famous New York natural spots, and they promoted sometimes fantastic ideas about past Earth history, based upon their isolated experiences. The waterfalls at Niagara had lured two centuries' worth of distinguished European visitors to travel through wild, Iroquois-occupied territories. Their widely circulated but often exaggerated reports reinforced the phenomenon's legendary impressiveness. At least eight different French explorers' accounts of Niagara Falls were published before the middle of the eighteenth century. Samuel de Champlain first called the world's attention to this sublime location by reporting what Native Americans told him about them in 1604. For the next seventy-five years, although some French-Canadian voyageurs and countless North American Indians could have provided eyewitness descriptions, none of the published European accounts of the falls were composed by people who had actually visited them.[3]

Father Louis Hennepin, a missionary accompanying Robert La Salle's 1678–79 expedition into the interior of the North American continent, was the first person who took advantage of the opportunity to publish firsthand observations of Niagara's wondrous natural phenomena. Unfortunately, his account was not accurate. Hennepin initially reported that the falls were approximately five hundred feet high (the vertical drop is in fact about 170 feet), and he expanded his exaggerated estimate to six hundred feet in revised editions of his memoirs.[4] Educated Europeans would not learn the truth for another seventy-five years; the report of a 1688 visit by the notorious Louis Armand de Lom D'Arce, the Baron de Lahontan, did nothing to correct Hennepin's excesses.[5] Peter Kalm, a student of Swedish botanist Carl Linnaeus, passed through the region aboard a boat in the summer of 1750, having first traveled on horseback from Albany to the British fort on Lake Ontario at Oswego. According to Kalm's analysis and measurements, the perpendicular drop of Niagara Falls was 173 feet, which led him to suspect that his predecessors were *menteurs* (liars); otherwise, one would have to maintain that all the terrain upstream from the falls had somehow mysteriously collapsed during the previous few decades.

Kalm's scrupulous eyewitness description of the surrounding topography and his assessments of the force and direction of the flowing water provided Europeans with some essential facts. His 24 August 1749 description of the gorge downstream from the falls, Queenston Heights, the aspect of the whirlpool, and the proximity of Goat Island to the head of the falls was more detailed than any previous report: "We first ascended Niagara River in a birch boat. The width of the stream here was said to be about twelve arpents, but this varied. The banks of the river consisted of high precipitous hills of red sandstone, in layers."[6] After describing an ever-present vapor, Kalm wrote that he and his French companions proceeded halfway to its source before stepping ashore to continue the expedition on foot. He explained, "It is difficult to come nearer [to the falls] with a boat, because [of] the number of steep rapids encountered. First we had to climb the high, steep river banks, then proceed three French miles [nine English miles] by land, which has two high and tolerably steep hills to be crossed."[7]

Concerning the falls themselves, Kalm presented a technically precise account of their dimensions and aspect, which would provide future observers with an invaluable baseline against which subsequent changes could be compared. Seated "where an enormous mass of water hurls itself perpendicularly down from a height of 135 French feet, i.e. 147 690/1000 Swedish feet, or about 34 2/3 fathoms," Kalm recorded in longhand exactly how the falls were situated with respect to the surrounding landmarks. As if describing a geometric object in Euclidean space, he continued: "The river at the falls runs from S.S.E. to N.N.W., and the falls themselves from southwest to northeast, not in a straight line but in the shape of a horseshoe or semicircle, since the island in the middle of the falls lies higher up than the ends of the semicircle. Above the falls and about in the middle of the river is an island, extending from S.S.E. to N.N.W., which is said to be about eight arpents long [about 1450 feet]." Remaining scrupulous about measurement and detail, he finished his description: "The island tapered in width toward its ends, being in the middle about a quarter of its length, with its lower extremity extending right up to the falls, so that no water tumbles down where the island touches the falls. The width of the latter, in its curved line, was said to be six arpents [about 1094 English feet]. The island lies exactly in the middle, so that the falls are split in two. The breadth of the end of the island in the falls is about 1/8 of an arpent [61 ft.] and the length of the island is, as I have just said, estimated by all to be eight arpents."[8]

Setting aside for a moment Kalm's attempts to convey a chorographic image in words, consider how his description grapples with the lack of a uniform system of measurement. Feet in France were not the same as feet in Sweden. English miles were one-third as long as miles in France. The standard arpent in Quebec was about 9/11 as long as a Parisian arpent, and this surveyors' term for river frontage was not even familiar outside of French-administered territories. Only *fathoms* enjoyed a universal relevance among seafaring nations, but it is clear from Kalm's description that even that term was not sufficiently precise to be used as a unit without qualification.

In the end, despite Kalm's focused attempt to report the configuration of Goat Island in exacting scientific detail, even he could not refrain from commenting on the spectacular impressiveness of the legendary place. Kalm closed his Niagara Falls narrative with these two extraordinarily proto-Romantic sentences: "It is enough to make the hair stand on end of any observer who may be sitting or standing close by, and who attentively watches such a large amount of water falling vertically over a ledge from such a height. The effect is awful, tremendous!"[9]

Over the ensuing decades, Europeans visited many places in upstate New York besides the famed waterfall and wrote more accounts full of hyperbolic description and geological speculation. For example, two French gentlemen traveling

Figure 1 Niagara Falls (from early print based on Peter Kalm's description). Unknown
artist, adapted from *Peter Kalm's Travels in North America* (London, 1770), edited by Adolph B.
Benson (New York: Wilson-Erickson, 1937; reprint, New York: Dover Books, 1987), 705.

in the 1790s, the Duc de la Rochefoucauld-Liancourt and the Comte de Volney, produced widely read but discrepant accounts of the natural phenomena
of North America, including the cataract at Niagara. Volney closed his analysis
of the waterfalls of New York, New Jersey, and Virginia with the bald assertion:
"To these grand phenomena of nature, Europe offers nothing worthy of comparison."[10] Such grandiose claims were naturally excerpted in popular American
newspapers and travel guides, and the public's appetite for travel became infused
with a growing fascination with geological matters. At the heart of all this interest lay one significant, fundamental question, How old is the Earth? Put another
way, can humans decipher the actual moment of Creation from observable evidence? Natural phenomena seemed to offer evidence that might help answer this
question.

American scientists of the early republic were not content to leave these majestic questions to be determined or resolved by foreigners. Benjamin Silliman, professor at Yale College and esteemed editor of the *American Journal of Science and
the Arts*, worked hard to develop an indigenous American scientific community
capable of taking charge of scholarly discourse and theoretical speculations about
American nature.[11] In 1821, Silliman published a letter he had received from
Amos Eaton under the title "The Globe had a beginning."[12] Having deduced
that rocks, gravel, and sand represent the result of ongoing erosive forces, Eaton
inferred that the Earth is not eternal. If continued indefinitely, he surmised, such

processes would level all mountains, leaving not one naked cliff; after sufficient time all land would be covered by "alluvial" soil. So, if Earth's history indeed had a beginning—whether or not it was as described in the scriptural Creation, a source of evidence virtually all early American natural scientists took seriously— the key would be to find a location where the processes of erosion were not only observable but measurable and delimited.

In 1822, the Rev. Edward Hitchcock, a theologian who began to develop a keen interest in geology after attending one of Eaton's popular lyceum lectures in Amherst, Massachusetts, responded to the New York naturalist's desire for loci where geological evidence could be traced back to the date of the Earth's creation. Hitchcock submitted a scientific paper proposing that the succession of annual winter ice damage to rocks adjoining the Connecticut River could be just such a "cosmogonical chronometer." Using this evidence, Hitchcock deduced that the Earth was relatively youthful: "Now every one must see that this levelling work cannot have been going on forever; and when we consider how very considerable is the quantity of rock yearly detached, and compare this with the whole amount of the debris, the conclusion forces itself upon us that the earth has not existed in its present form from eternity." Admitting that the technique was too imprecise to absolutely date Earth's creation, Hitchcock nevertheless confidently concluded "that Moses placed the date of the creation too far back, rather than not far enough."[13]

Because Peter Kalm's description offered such an accurate and reputable baseline, however, students of the geology of the Niagara gorge insisted that the falls might be a far better place to calibrate the passage of terrestrial time. The erosive power of the massive quantity of water flowing over that precipice, when considered together with the dramatic cliff faces that surround a deep trench along both sides of the river downstream for several miles, seemed to offer an empirical method for determining how far and how fast the falls were cutting through the layers of rock. The shifting position of the waterfall, in other words, traced a pathway of gradual recession from a beginning point in time that eighteenth-century observers were quick to link with the Earth's creation. This idea assumed, of course, that one did not worry about the time or circumstances required to deposit the layers of red sandstone embedded in those cliffs from before the excavation of the gorge began. Taking such assumptions as givens, two technical questions remained to be resolved: Have the falls receded at a constant pace over all time? Which location between the present site of the falls and the place where the Niagara River empties into Lake Ontario was the starting point of the erosion? These two questions formed the basis of an active line of geological research and

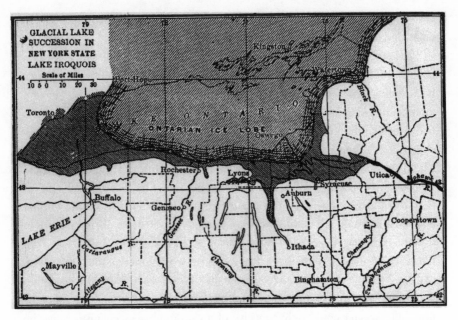

Figure 2 Glacial Lake Iroquois. In 1911 Isaiah Bowman—then a Yale professor, later director of the American Geographical Society and president of Johns Hopkins University—depicted the Ridge Road as Lake Ontario's former shoreline, hinting at a "lake bursting through mountains" scenario. (The Mohawk River valley would have provided a drainage route to the Atlantic Ocean.) Bowman, *Forest Physiography* (New York: John Wiley and Sons, 1911), 713. Image available at http://etc.usf.edu/maps/pages/1000/1053/1053.htm.

debate in the early nineteenth century, and New York State became an important destination for those who wanted to pursue such important scientific questions.

Beginning with the records compiled by post–Revolutionary War township surveyor Andrew Ellicott in 1790, geologists attempted to build a collection of longitudinal data from which it might be possible to calculate rates of recession. Ellicott's record of official measurements along the Niagara River reported a 162-foot drop at the "Great Fall." As Peter Kalm had done forty years earlier, Ellicott also sent a more complete description to a colleague to be published in a European journal.[14] Comparing the distances from shore landmarks to the falls, as described in just Kalm's and Ellicott's accounts, contemporary theorists offered rough estimates of the rate of recession ranging from about one foot to a little more than two feet per year. If the Niagara had scoured the entire gorge along Queenston Heights from its northernmost point to the current location of the falls, and if that erosion had occurred continuously at these rates since the

Creation, participants in this debate calculated that the Earth might be between seven and seventeen thousand years old.

Some observers, however, noted that the Ridge Road paralleling Lake Ontario's shore further complicated the problem; if the present lake level had once been higher, this would affect the cumulative drop (and therefore the erosive force) of the waterfall. Caleb Atwater, an amateur geologist and antiquarian who had relocated from central New York to Ohio in the 1810s, published communications to the *American Journal of Science* in 1818 suggesting that Lakes Erie and Michigan may once have been merged with Lake Huron. He imagined these individually impressive bodies of water together might have comprised a gigantic Great Lake, submerging major portions of now dry land in northern Ohio and Indiana and much of the Michigan territory. Atwater also offered reasons to suppose that Niagara's erosion chronometer might postdate the history of Earth. Evidence of prehistoric drainage patterns suggested to him that the Great Lakes had at some time emptied southward into the Ohio and Mississippi Rivers, rather than northeastward via the Niagara and Saint Lawrence Rivers. To account for this shift, Atwater could only conceive of a catastrophic mechanism, since he (like all contemporary theorists) was unaware of the agency of ice: "Lake Ontario, for some cause (possibly an earthquake, or the wearing away of its outlet, or both) is considerably lower than it was formerly: in that way the land along its banks, once covered by its waters, is drained, presenting appearances exactly similar to those seen in many of our prairies [in Ohio]."[15] New York's most famous natural attractions were coming to be appreciated as windows into the mysteries of Earth's geological history, but the task of reading that history was proving to be very problematic.

MOHAWK VALLEY: EVIDENCE OF LAKES BURSTING THROUGH MOUNTAINS?

Americans imagined their continent to have been a geologically volatile place. Perhaps this idea was an unconscious reflection of their own pride in having waged a successful political rebellion, but in any case early travelers' reports provide ample evidence for the widespread belief that past geologic revolutions of land and sea had substantially altered the landscape since Creation. A series of letters written by Yale's highly respected president Timothy Dwight, for example, suggest that the land now inhabited by the citizens of the new nation had been sculpted purposefully, as if a deity had anticipated their needs, through a combination of ordinary erosion processes and extraordinary natural calamities. This New England Congregationalist theologian integrated a highly developed sense

of scientific curiosity into his secular writings about the landscape and the inhabitants of New England and New York.[16] Like the Deist Thomas Jefferson, Dwight
saw tremendous value in the study of nature for the moral instruction of the new
generation of Americans, and the intellectual and amateur scientific reach of
Dwight's letters certainly approached the standard established by Jefferson's 1785
Notes on the State of Virginia.[17] In his letters, Dwight freely speculated about
Earth's history based on what he knew from the Bible and what he saw in the
landscape. His letters sample the natural history ideas that prevailed in those early
years when Benjamin Silliman, Dwight's protégé in New Haven, became the
republic's best-known professor of natural sciences.

Reverend Dwight came to New York State predisposed to see in its scenery
evidence supporting the biblical account of the Flood. On one of his earliest
trips to central New York in 1799, Dwight imagined scenes radically different
from the tranquil, humble farmsteads sprinkled before him on the banks of the
Mohawk. Passing Little Falls, Dwight deduced, "The hills on both sides are steep,
and ragged, and strike the eye at a glance, as if they formerly were united, and,
thus, presenting a barrier to the waters of the Mohawk, converted them into a
large lake, which covered all of the low grounds, as far back as the hills West of
Whitesborough." Observing that "the rocks exhibit the most evident proofs of
having been formerly worn and washed, for a long period," Dwight found further confirmation for his theory of violent inundation as he proceeded westward,
where he thought the hills looked like banks and the stones resembled "those,
which are found at the bottom of lakes and rivers."[18]

A subsequent trip north of Albany convinced Dwight that Lake George owed
its existence to a similar catastrophe, perhaps related to the opening of that gap
in the Mohawk valley at Little Falls. He began his description of Lake George
by reciting a quote attributed to Moses: "The fountains of the great deep were
broken up." Drawing upon the authority of Christian scripture where it seemed
most relevant to the subject, Dwight modestly claimed: "To my eye, at least, the
general aspect of the whole scene; the appearance of the strata; the forms of the
mountains; the manner, in which they descend to the lake; the figures, presented
by their several points; the continuation of those points under the water; the manner in which they are connected with the islands; the appearance of the islands
themselves, the surfaces, and strata, of which in many instances are horizontal,
and where they are oblique, have their obliquity easily explained by their reference to the nearby shores; all evinced this fact so strongly, as to leave in my mind
not a serious doubt."[19] The technical language, cautiously speculative tone, and
evidence of thorough investigation (claiming to have traced specific rock layers

beneath the surface of the lake!) demonstrate how well versed the Yale president had become in mimicking the form and style of contemporary geological discourse.

It was one thing to report one's observations but quite another to assemble evidence to support a theory. Believing that valleys were formed because lakes burst through mountains, Dwight revisited the waters flowing past Little Falls in 1804, reaffirming his impression that there must formerly have been a lake there: "I am entirely satisfied, that the mountains, which ascend immediately from the banks of the river, were anciently united; and that the river above formed an extensive lake, gradually emptied by the wearing away of the earth and stones, which originally filled the gap. On the rocks, bordering the road, unequivocal marks of the efficacy of water are visible, at different heights, to forty feet above the level of the road, and to fifty above the river." Returning to the same spot yet again in 1809, Dwight reiterated his speculative scenario: "Of course, the waters of the Mohawk found a barrier at the Little Falls, more than one hundred feet in height, and were therefore certainly a lake extending far back into the interiour." But this time he extended his hypothesis, arguing from the suggestive evidence at Little Falls that a rush of water must at one time have scoured the entire Mohawk valley: "In one case, then we are faced with the demonstration, so far as reasoning from facts may be called such, that the waters of a river, which has now washed away its barrier, were anciently confined by the jutting of mountains so as to constitute a large lake, agreeably to the scheme mentioned in the account given in my second journey to Lake George. Fair analogy will convince an observing traveller, that there were once lakes of the same class in all the places which I have specified, and in many others."[20]

But if such an ancient catastrophe had occurred in the uplands west of Schenectady, then what might have happened further downstream? Dwight began to ponder the geological origins of the wide Hudson River valley: "It is a remarkable fact, that the Hudson should have found so fine, and safe, a bed, in a country so rough, and between banks so often formed of mountains, or high hills, and to so great an extent abutting it in precipices of a stupendous height. Yet even through the highlands its navigation is perfectly uninterrupted." He blended his theological and scientific beliefs to account for this unusually wondrous and useful natural avenue. Borrowing from his earlier observations, he now proposed a familiar scenario for how the Hudson valley was formed: "The country North of the highlands, from Fishkill to Waterford and possibly farther still, was, as I believe, and as I have heretofore observed, once a vast lake. The valley of the Hudson is here in some places not far from forty miles in breadth. The mountains on both sides form a complete barrier to the waters of such a lake. On the South the highlands effectually kept up these waters to a great height, not improbably for a long period after the deluge."[21]

Dwight understood that his "lakes bursting through mountains" mechanism was logistically complicated. He tried to analyze the geological dynamics of a deluge. "If the channel between Butter Hill and Brecknock, at the entrance of the Hudson into the highlands, was worn out suddenly, it was probably accomplished in a manner, resembling that, in which the lake in Glover, elsewhere mentioned in these letters, forced a passage for its waters, two hundred feet in breadth and depth, within the limits of twenty-four hours." What kinds of enduring consequences could be traced back to such a transitory flood event? Dwight speculated, "The surface of the earth, surrounding this lake, was hard, but the inferiour strata were, to a great depth, light and loose." These lightweight rocks would have eroded easily and rapidly in the turbulence of a torrent. He concluded, "If we suppose the Hudson a lake at any given, ancient period; the efflux of such an immense mass of waters must go far towards explaining the great depth of its present channel."[22]

The good reverend did not rest complacently upon his conclusion. Inspired by the awesome power of water New Englanders were beginning to experience in the suddenly deforested mountain valleys of Vermont and New Hampshire, where tiny mountain freshets occasionally raged into disastrous spring torrents, Dwight extended his geological theory to the furthest reaches of the state, and beyond. "It has been supposed, that Long-Island was once a part of the Continent; and that by a great convulsion, which I have mentioned, it was separated from it by the intervention of the sound." He supposed, however, that this "intervention" must predate Noah's flood "because there are no traces [on Long Island] of any channels, worn by the rivers which lie Westward of the Connecticut; particularly the Hooestennuc [Housatonic]."[23]

One did not have to be a Christian theologian to entertain such cataclysmic visions of earth history in North America. Thomas Jefferson described the passage of the Potomac and Shenandoah Rivers through Harpers Ferry in similar terms: "In the moment of their junction [the two rivers] rush together against the mountain, rend it asunder, and pass off to the sea." Jefferson elaborated the chain of events as follows: "that the mountains were formed first, that the rivers began to flow afterwards, that in this place particularly they have been dammed up by the Blue ridge of mountains, and have formed an ocean which filled the whole valley; that continuing to rise they have at length broken over this spot, and have torn the mountain down from its summit to its base."[24] Jefferson thus had his own lake bursting through mountains. Other geological writers would reiterate this idea to explain a variety of dramatic river valleys ranging from narrow passages along the Schuylkill and Susquehanna Rivers, to the Delaware Water Gap, to West Point.[25]

The ripple effect of contemporary scientific discussions carried ideas about "lakes bursting through mountains" into more popular writing, such as the published travels of William Darby. Darby first visited the vale of the Mohawk in May 1818, bringing along Samuel Latham Mitchill's edition of French geologist Georges Cuvier's *Theory of the Earth*.[26] In his book, Darby signaled his expectation in advance that a formerly extant great lake across central New York must have created the passage at Little Falls. Surprisingly, Darby cited neither Mitchill nor Dwight as an authority for this interpretation but rather DeWitt Clinton, then sitting governor of the state. He predicted that Clinton's description "of the Little Falls, will continue to supersede the use of any other. It is indeed a fine specimen of topographical painting, and places the attendant phenomena before the mind's eye. . . . [A]mid the wildness of the scene, and in hearing of the roar of the gushing waters, [I] read and felt the truth of this excellent view of one of the great scenes that our country presents to the admiring traveller."[27]

Darby proceeded finally to indulge his own speculations, not only drawing upon the familiar "lake bursting through mountains" hypothesis but also elaborating inventively (and presciently, in terms of later glacial explanations) upon the historical drainage patterns of upstate New York: "When writing on the subject, the idea first presented itself to my mind, that through either the Mohawk, or some valley to the southwest of [Rome, New York] once flowed the St. Lawrence river [i.e., the outlet to Lake Ontario]. Rome is only 188 feet above lake Ontario." Darby surmised: "If ever lake Ontario was 188 feet above its present level, then was its waters discharged, either towards the Hudson or Susquehanna, or both. The ancient features of this continent, must have been very different from the present." The broader implications were there for anyone to deduce, in Darby's opinion: "No person of ordinary observation, who examines the shores of lake Ontario or the banks of St. Lawrence River, but will grant that abundant evidences remain to attest an elevation of lake Ontario of more than two hundred feet above its present surface. Evidences also exist to prove that the recession of that lake was periodical. The chain of smaller [Finger] lakes lying west of Rome, north of the dividing ridge [between the St. Lawrence and Susquehanna watersheds], and east of the Genesee river, were once bays of Ontario.[28]

Travel accounts like Darby's contributed to a growing popular discourse on New York geology which drew liberally from early published scientific descriptions such as those offered by Mitchill and Clinton. Through these narratives and the commentaries provided by travel writers, Americans learned more about the great French naturalist Georges Cuvier and his cutting-edge ideas about catastrophic geological change; he posited that exchanges of sea and land were responsible for marine creatures' fossils being found far inland. Through their readings of Cuvier in translation, Americans also learned more about his Enlightenment predecessors in French natural history and natural philosophy, such as the monumental *Histoire Naturelle* by Georges Louis LeClerc, Comte de Buffon. The late eighteenth-century theories of James Hutton, the Scottish farmer who championed the idea of volcanic formation of all rocks, and Abraham Gottlob Werner, the German mineralogist from Saxony who taught that all rocks were deposited through the action of water (ranging from chemical precipitation of sediments out of solution to the effects of violent flooding), were already widely known among English readers of geological ideas in the early nineteenth century.

The American reading public became increasingly intrigued by geological knowledge between 1800 and 1840. "Learned" travelers were clearly so common, so fanciful in their pronouncements, and yet so serious in their demeanor as to be a legitimate subject of farce. In the anonymously published parody *A Knickerbocker Tour of New York State*, geological tourists were caricatured using the comic persona first established in Washington Irving's *A History of New-York* (1809): "Read Mitchell's [*sic*] notes, as to the region through which the Mohawk flows; will instruct you amazingly, haven't read them as yet, but purpose so to do, when we have leisure and inclination." If you want to speak with authority about what you are seeing out the windows of your coach or packet boat, the satirist suggests, there are experts to be consulted. And yet, a democratic society like ours offers no obstacle to voicing an opinion, however uninformed. At its most absurd, this was a literature in which vacuous metaphor masqueraded as vivid description: "Hills lofty and rounding, much the appearance of waves running 'mountains high,' as the phrase is." On a gloomy day with clouds threatening rain, the weary traveler might seek comfort in a good book. Unfortunately, the satirist reported finding nothing on board his boat but religious tracts and a worn out copy of Hoyle's games, so he returned to enjoy the scenery from beneath an umbrella on deck: "Thought of the deluge, which led to a 'lucubration' on the regions we were then passing through. The singular 'aspect of the contour' of the hills, tends to support the theory of Mons. Buffon, who supposes the earth (when in a state of 'Chaos') to have been a huge mass of heated matter, rolling about in confusion in a fluid state, being nothing 'more or less' than an immense quantity of batter, or for instance, very like an enormous Indian pudding!" From this humble simile, the Knickerbocker reached back up into the realm of the spectacular, his language again in keeping with the standards of popular historical geology, the new American literary genre Mitchill had already done so much to establish: "It (the earth), having been descended from the Sun, by the conjunction of a Comet; and as the creation proceeded, 'all the various bodies and particles of this heterogeneous mass cooled, and took their places as we now find the different strata,' thus leaving the world in a proper manner constructed, a most excellent piece of work."[29]

Whether the object of satire or serious investigation, New York's wealth of natural marvels and geological puzzles continued to entertain and perplex people traveling between the Atlantic seaboard and the continental interior. The attractive, intriguing landscape had been shaped by natural forces so complex as to make the task of geological investigation far from easy.

CHALLENGES AND OBSTACLES TO INVESTIGATION

In the late eighteenth century, overland travel in the United States was notoriously arduous. Mounted on horseback, a traveler might make good time and skirt natural obstacles with skill and foresight, but a passenger aboard any vehicle obligated to follow a road was likely to endure all kinds of delay and physical torment. Rhode Island historians Hadassah Davis and Natalie Robinson describe why a trip from Boston to New York along the relatively developed post roads ordinarily took a week: "These roads were little more than dirt tracks worn by cart or coach wheels."[30] Travel through New York State's interior was even more challenging, aside from well-worn paths that hugged the shores of the Hudson and Mohawk Rivers. Isolated hills or extended ridges forced people to either go the long way around or surmount and descend elevations along winding hairpin paths. Weather could exacerbate the difficulties of travel even across level terrain. According to Timothy Dwight, the roads he took westward beyond the navigable parts of the Mohawk River in the summer of 1799 were "almost always soft; in moist seasons a mass of mud; and in wet seasons intolerable." He concluded that the experience of traveling was "not merely uncomfortable and discouraging, but an Herculean labor."[31]

Geologically inclined passengers had many lingering opportunities to examine the roadside, but unfortunately subterranean structures were rarely exposed to view. Only a stream crossing here and there, or the occasional catastrophic collapse of an eroded, perhaps deforested hillside wherein bedrock outcrops and layers of strata might be seen from the road, might open a window onto the Earth's past. After all, eighteenth-century roads were not cut into hillsides like the highways produced by modern road engineers. Such an effort would have entailed an egregious expenditure of labor, given the speed and design of contemporary animal-powered vehicles. Traveling along an interstate highway today, one can contemplate "all the sheets of rock that had been bent, tortured, folded, faulted, and crumpled."[32] Modern roadways inspire geological thinking by providing sustained visual access into the dynamics of rock formation. Roads in 1800 inspired only backaches, frustration, and anxiety.

In the absence of a developed transportation infrastructure, the serious geologist had to become an accomplished bushwhacker to be able to say anything about the distribution of rock types beneath the soil. Despite the vast advantages road and canal construction provided by 1824, Amos Eaton complained that the geologist still had to go to extraordinary lengths to venture even a shaky guess about the underlying stratigraphy of a territory. To overcome the fact that 99 percent of the desired information was hidden from view, one was "compelled to traverse every stream, to search out every naked cliff, to descend into every cavern, to examine wells, water raceways and ditches, [just] to obtain a few elementary facts."[33]

The geological promise of New York's magnificent natural phenomena was clear enough to early nineteenth-century observers, but the fulfillment of this promise was going to require substantial investment of material social, economic, and political energies. New ideas and new structures would also have to emerge. For a state's government to take an active role in harvesting its natural bounty, public awareness would have to be guided by a variety of unprecedented activities. Pressure to improve roads and support innovations in alternative transportation technologies (steam boats, canals, and eventually railroads) grew across the American republic in the early nineteenth century, and New York would take the lead in pioneering the material and social mechanisms needed to accomplish these changes.

CHAPTER TWO

Natural Sciences
and Civic Virtues

This discovery of Clinton, therefore, although hardly noticed by his countrymen, procured him much reputation among the learned of Europe; and the diplomas of many societies founded for the cultivation of natural science were showered upon him.
— *James Renwick (Columbia College Professor of Natural Experimental Philosophy and Chemistry), 1840*

HONOR IN THE EARLY REPUBLIC

At dawn on the morning of 11 July 1804, America's third vice president shot and mortally wounded the nation's first secretary of the treasury, thus ending two great New Yorkers' political careers. At the same time, incidentally, the clear way forward for a third opened. DeWitt Clinton built his career using the same skill, ruthlessness, intelligence, and maneuvering as his rivals Aaron Burr and Alexander Hamilton, but he also pursued scientific achievement. Although Clinton had once participated in a bloody (though nonfatal) duel, exchanges of gunfire to defend one's political reputation lost credibility after Hamilton's death, and it was Clinton's exemplary activity as a public intellectual that cemented a new way to exercise and display civic virtue in American political culture. Contemporary observers understood how relevant the study of geology would be to the success of a venture so ambitious as the Erie Canal, and it is strange how little attention modern historians have paid to Clinton's involvement in the natural sciences, given that the canal is universally cited as the centerpiece to his political career.[1]

Following the American Revolution, all kinds of new social patterns needed to be invented. The success of the rebellion raised doubts about many forms of authority that had been familiar but unchallenged before. Although Europe's absolute monarchies provided the only kind of sovereign governance structures that were known to work, aristocratic titles and deferential manners did not fit with the new ideology of a revolutionary democratic republic. England offered the example of a bicameral representative legislature, and the designers of the American Constitution adopted this model as a framework for a federal congress in the late 1780s. Parliament's two houses were clearly distinguished by social class; landed nobility populated the House of Lords, while literate artisan, commercial, and industrial elites were represented in the House of Commons. Without an obvious class system, how would citizens of the United States be deemed worthy to serve in the upper house as a senator rather than in the lower house as a congressman? A wealthy American who wanted to secure a powerful political reputation in the early republic needed to find ways to exemplify positive aristocratic qualities without slipping into the arrogance or false pretensions of corrupt entitlement.

The struggle to invent and embody new forms of government revolved around how to cultivate personal honor outside the sociopolitical structures of monarchy. In a democracy, where "reputation was a matter of public opinion," all kinds of personal disputes spilled over into public political discourse.[2] Even before political party rivalries became well established, campaigning relied upon the effectiveness of gossip, slander, and rumors to a degree that might startle anyone who imagines mudslinging to be a modern invention. Writers using pseudonyms routinely published damaging allegations and counterarguments. Ugly lawsuits, and even the occasional trial by combat, ensued when political leaders found themselves embroiled in questions of honor. As one historian points out, "Certain slurs were off limits, tame as they are by modern standards. *Rascal, scoundrel, liar, coward*, and *puppy*: these were fighting words, and anyone who hurled them at an opponent was risking his life."[3] Beyond these negative aspects of the problem of honor, the question remains: How did aspiring American politicians positively cultivate honor? Military service was clearly a popular option, especially for the generation that had served in the Revolutionary War. For others, demonstrations of intellectual prowess offered an alternative pathway to prestige and respect.

The rise of party politics in New York provoked all kinds of anxieties as the eighteenth century raced to a close. Political ambitions were exercised within a highly volatile universe of social, economic, and personal propriety. The circumstances of these early contests for power and prestige were neither pretty nor always fruitful. Two politically significant duels were fought atop the stark cliffs that overlook New York City from across the Hudson River: the one conducted in the fall of 1802 is now generally forgotten, while the other, in the summer of 1804, is notorious for its fatal consequences. Both bouts highlight the risks involved in cultivating a positive public reputation in the early years of the American republic. DeWitt Clinton's ascent to become New York's preeminent public figure in the first quarter of the nineteenth century was a consequence of those bullets fired at Weehawken, New Jersey.

JEFFERSON AND ALL THE KING'S MEN

DeWitt Clinton was a Jeffersonian Democrat-Republican at a time when New York and New England were dominated by Federalists, the original party of American patriotic nationalism. Many of New York's elite politicians were prominent leaders in this party, which had established the architecture of the federal Constitution and secured the success of George Washington's presidency. Like Clinton, many of his state's leading Federalists had burnished their intellects at King's College, colonial New York's premier institution of higher education. It is worth remarking just how unusual such an education was for the first generation of American citizens. Roughly one out of every two thousand free men would have attended one of the nine colleges operating in colonial America: Harvard, Yale, William and Mary, the University of Pennsylvania, the College of New Jersey (Princeton), Brown, Queens College (Rutgers), Dartmouth, and King's.[4] Of these, King's was known as a bastion of classical learning and conservative political ideology, though Loyalism evidently failed to make a lasting impression upon the many brilliant rebels who attended the school. Once the war for independence ended, King's would be renamed Columbia College in honor of a famous Italian (deliberately not an English!) navigator.

DeWitt Clinton's education at King's College began while the last remaining British troops still occupied New York City and the future of American independence was in the hands of Paris negotiators. Too young to partake in military combat, Clinton instead was given the opportunity to develop his understanding of European languages and the natural sciences. From 1784 to 1786, the college hired Dr. Henry Moyes, a popular lecturer who had been blinded by smallpox at the age of three, to be its first professor of natural history and chemistry. Hailing from Scotland, where he socialized with Adam Smith and other leading members of the Lunar Society, Moyes brought to New York the latest debates in European chemistry. Though Moyes was an early champion of Joseph Priestley's phlogiston theory, the lectures Clinton attended were probably the first in North America to present Antoine Laurent Lavoisier's as-yet-untranslated combustion experiments, which would ultimately recognize and redefine Priestley's "dephlogisticated air" as the elementary substance we now know as oxygen.

King's College offered an even more modest natural science curriculum back in 1768, when New York Federalist Gouverneur Morris graduated at the precocious age of sixteen.[5] As a child, Morris had displayed an unusual knack for solving complex mathematics problems in his head. His mathematic capability would be useful in the early 1780s when he was given the opportunity to invent and implement an entire monetary infrastructure from scratch and came up with the American decimal currency system. Before that, however, in 1771, Morris earned a master's degree in the traditional literary classics at King's while simultaneously pursuing private legal studies. Morris was admitted to the New York bar before he turned twenty. With the outbreak of the Revolutionary War, this multitalented scholar rapidly became a leader. Morris was active on the Committees of Correspondence, Safety, and "Intestine Enemies" until British armies occupied the city, and despite his youth he was chosen to represent the colony at the Continental Congress in Philadelphia. Eloquent and succinct, in 1787 Morris authored the stirring phrases that form the Preamble of the Constitution: "We the people of the United States, in order to form a more perfect Union . . . " An unabashedly aristocratic elitist, Morris became one of the strongest proponents of higher education and meritocracy among the Founding Fathers, insisting that "the United States would need to rely on its most educated citizens primarily in its governmental affairs."[6]

Two other important New York Federalists, Alexander Hamilton and John Jay, were also King's College alumni. The Caribbean-born Hamilton, five years younger than Morris but a dozen years older than Clinton, entered King's in 1774. His education prepared him for an illustrious, daring, and accomplished career as an economist and political philosopher. After the Constitution was written in 1787, Hamilton initiated and promoted a brilliant, erudite public relations campaign urging its ratification; he disseminated the series of anonymously published essays now known collectively as the Federalist Papers. At Hamilton's invitation, the venerable lawyer and diplomat John Jay (class of 1764) contributed a few essential essays containing arguments that laid the legal and ideological groundwork for sewing the original thirteen states into the United States. Serving as George Washington's secretary of the treasury, Hamilton went on to study British models of central government power over finance. His creative and bold adaptations of those models enabled the newly independent nation to stabilize its massive debts, establish the first national bank, and proceed to embrace and actively participate in industrial development and international trade. For his part, John Jay served as the nation's first chief justice of the Supreme Court and negotiated the controversial treaty that averted the threat of renewed warfare with Great Britain in 1794.

For the duration of George Washington's presidency, Federalism functioned more or less as a unifying national political ideology. In the wake of his retirement from public life, however, longstanding tensions among the strong-minded policy-makers who had served in his cabinet spilled over, giving life to a much-dreaded reign of political "factions." In 1797 Thomas Jefferson, a Virginian educated at the College of William and Mary, became the nation's second vice president. Having served as a congressman, diplomat to France, and the first U.S. secretary of state, Vice President Jefferson advocated for a democratic counterbalance to what he saw as the monarchical tendencies Jay, Hamilton, and John Adams had shown when they served together under President Washington. Now Jefferson emerged as champion of the Republican critique to Federalism. He would vainly strive to balance his conflicting roles of second in command and chief voice of opposition throughout the one-term Adams presidency.

With Benjamin Franklin's death in 1790, the United States lost its first world-renowned man of science. In this regard, Jefferson was Franklin's heir apparent. The membership of Philadelphia's American Philosophical Society tapped the Virginian to serve as its president in 1797. He celebrated that occasion by present-ing a technical paleontology paper, in which he described several fossils recently found in western Virginia. Jefferson postulated a large quadruped "megalonyx" (giant claw) in the belief that the fossils belonged to the family of large cats.[7] It was a highly regarded piece of research, and Jefferson valued the position of leadership he had earned among respected colleagues such as the transplanted Englishman Joseph Priestley more than his lofty elected public office.

Jefferson had little reason to imagine that his successful pursuit of scientific respectability would enhance his political standing. Up to that time, his experi-ence had been quite the opposite. A natural history tour through New York and New England in the summer of 1791, undertaken with fellow politician James Madison and the botanist John Beckley, was a rare chance for Jefferson to satisfy his curiosity about the natural phenomena of this region, and his desire to study the damaging effect of the Hessian fly on American agriculture provided a practical focus for their inquiries. Suspicious Federalist politicians regarded the expedition as a ruse, however. They gossiped privately and intimated publicly that Jefferson and Madison's trip was no innocent ramble in search of curious rocks, animals, and plants. Rather, these prominent Virginians were presumed to be poking around rural northern communities in order to spread the ideas of French revolutionary extremism, drum up opposition to Alexander Hamilton's ambitious banking plan, and incite general agitation for renewed hostility toward Great Britain. Perhaps the secretary of state was even hoping to organize Shaysite rebels to regroup![8] Unfortunately, innuendo like this could inflict a kind of political damage that no amount of intellectual achievement would undo.

Two years later in January 1793, Jefferson, then vice president of the American Philosophical Society announced the society's sponsorship of a journey to "explore the country along the Missouri [River] to the Pacific," to be led by the famous French botanist André Michaux. However, with the arrival later that spring of Edmond Genet, revolutionary France's overzealous new minister to the United States, the expedition's purpose underwent a troubling transformation. Genet informed the secretary of state in early July that Citizen Michaux would now carry instructions from the French government as well as the American Philosophical Society. According to Jefferson's private memoir, the *Anas*, Genet's plans for the expedition involved little botany but instead outlined how insurrections were to be sparked in British Canada and Spanish Louisiana. Within two weeks, Michaux was on his way west intending to raise a force of armed Americans to liberate New Orleans. Genet's plans to destabilize France's enemies were foiled when the retired American general James Wilkinson, secretly an agent of the Spanish government, sent word from Lexington, Kentucky to his employers that fellow Revolutionary War general George Rogers Clark was brazenly recruiting fifteen hundred brave volunteers to "take the whole of Louisiana for France."[9] Consequently, Jefferson's hopes for a transcontinental voyage of discovery were put on hold indefinitely.

An opportunity to redeem the relevance of scientific achievement to American politics arose during the hotly contested election of 1800. A Pennsylvania Federalist published a pamphlet attacking Jefferson's interest in natural philosophy, specifically arguing that the love of natural science rendered a man incapable of being an effective political leader. In an appeal intended to arouse anti-intellectual tendencies among the electorate, the anonymous author rationalized: "Science and government are different paths. He that walks in one becomes, at every step, less qualified to walk with steadfastness or vigor in the other."[10] This attack was reproduced widely in the press. From Kentucky came one enlightened Republican's response: "The philosopher is nothing more than a being . . . whose opinions are drawn from the convictions of truth and reason." This supporter retorted that Jefferson's practice of scientific reflection "had given to his mind a degree of philosophical tranquility infinitely superior to most of his contemporaries."[11] While the outcome of the election surely was not decided on this issue, no one could dispute that the Republican standard bearer was unabashedly intellectual. Jefferson's strong commitment to the cultivation and the diffusion of scientific knowledge in every aspect of his public life helped to promote the general respectability of science within the broader American culture.

Indeed, Jefferson's cerebral reputation remains legendary to this day. Generations of historians, struck by his interest in natural history and natural philosophy have magnified the legend by endlessly reinforcing these aspects of Jefferson's genius.[12] Jefferson actively participated in contemporary scientific disputes with European naturalists and philosophers, most notably the debate with Georges Louis Leclerc Buffon regarding geographical determinism and the robustness of New World species. Like Franklin, Jefferson learned how to elevate his domestic political reputation through the reflected prestige of international scientific celebrity. Unlike Franklin, however, Jefferson enjoyed the opportunity to bring this experience directly into play as the nation's chief executive. As a result, the third president was able to imagine and articulate a vision of nature that linked a tradition of publicly funded American scientific expeditions to a policy of expanding national power through territorial acquisition.

The Lewis and Clark Expedition of 1804–1806 was more than just the fulfillment of the dream dashed a decade earlier when Genet hijacked the Michaux expedition. Geopolitical circumstances had changed radically, and the United States was now enduring the troubling effects of the global Napoleonic wars between Great Britain and France. The 1803 Louisiana Purchase was a bold diplomatic move that had suddenly doubled the size of the country but also risked overextending the power of a small federal government. Jefferson responded to the challenge by systematically equipping his private secretary Meriwether Lewis with the material and intellectual wherewithal to make meaningful Enlightenment-era scientific discoveries. Charged with a diplomatic (not a military) assignment, expedition coleader Lt. William Clark (George Rogers Clark's younger brother) was instructed to seek to establish friendly relations with all the native groups they encountered while gathering basic geographical intelligence. In addition to considering the salient political benefits of such exploration, Jefferson fretted minutely about the science. He tutored Lewis privately in botany, zoology, and geology so that the Corps of Discovery might bring back not only sketches, maps, journals, and trinkets but also a collection of properly documented specimens of native flora, fauna, and minerals. In effect, Lewis and Clark were dispatched to begin the process of taking inventory of the North American continental interior, a vast and varied set of contiguous regions that would require decades, if not a century, for American citizens to fully occupy and exploit.

Contemporary observers understood just how central the study of nature was to the success of Jefferson's domestic policies and his international reputation. Eulogizing Jefferson before the New York Lyceum of Natural History in 1826, Dr. Samuel Latham Mitchill described how natural science and politics formed complementary halves of Jefferson's whole genius. Mitchill recalled in detail how the president took a great interest in and derived advantage from the success of the Lewis and Clark expedition. In 1807, Jefferson assigned the returned Captain Clark the task of digging up a wealth of fossil specimens at Big Bone Lick in Kentucky, Clark's home state. During the spring of 1808, citizens of the nation's capital were treated to an extraordinary display of the massive tusks, bones, teeth, skulls, and vertebrae of bison, elephants, and mastodons. Jefferson divided these fossils into three equal parcels: one to become a part of his own private collection, one to go to the American Philosophical Society in Philadelphia, and one to be donated to the National Institute of France. Mitchill thought it pertinent to remark that the scientific elite of Paris were so impressed that they elected Jefferson to the "exalted situation of a foreign member of the national institute" in respect and gratitude for these valuable additions to their paleontology materials.[13]

While few if any subsequent residents of the White House have matched Jefferson's extraordinary breadth of scientific curiosity and ambition, it would be a mistake to see him as a solitary bright light in an otherwise dim universe of early republican politics. The intellectually astute Morris served in the Senate from 1800 to 1803, and though he was succeeded by a scattering of unremarkable Federalists who were Revolutionary War veterans, New York was also represented by DeWitt Clinton (1802–1803) and Mitchill (1804–1809). Besides their affiliation with Jefferson's political views, Clinton and Mitchill were close friends who shared such strong interests in all aspects of nature that they could hold their own in philosophical conversations with the president.

Mitchill had received a more impressive education in the natural sciences than any of the King's men. Sent by his family to study medicine at the University of Edinburgh, he studied several branches of natural philosophy at the epicenter of the Scottish Enlightenment.[14] Besides the medical lectures offered by eminent chemist and physician Joseph Cullen, Edinburgh's intellectual community at this time was home to a wide array of groundbreakers, including Adam Smith in economics, Joseph Black in physics, James Watt in the engineering of steam power, John Playfair in mathematics, and James Hutton in geology. Mitchill could have crossed paths with any of these men, though he arrived too late to encounter the philosopher David Hume (as Henry Moyes had). Upon his graduation, Mitchill returned to New York equipped to serve as a distinguished physician and professor. He taught chemistry, botany, zoology, and mineralogy at Columbia from 1792 until 1801 while acting intermittently as a legislator in the state assembly. In the notorious election of 1800, Mitchill won a seat in the U.S. House of Representatives, an honor that temporarily interrupted his scholarly career. Still holding the seat he gained in the Senate in 1804, Mitchill resumed teaching at the College of Physicians and Surgeons in New York in 1807. In later years, he also helped to found Rutgers Medical College in New Jersey.

With refined intellectual political leaders like Morris, Clinton, and Mitchill in office, the natural sciences in early nineteenth-century New York enjoyed a kind of bipartisan support. All three men were generous promoters of education, powerful patrons for research and scientific collecting activities, and learned and able practitioners of natural history. To a degree frequently understated by modern historians, natural science provided the key to Clinton's revolutionary canal-era innovations in culture, politics, commerce, and technology. The relationships between his prominence as the first successful prophet of internal improvements, his experiences in the political leadership of the country during the early national period, and his lifelong championship of science in public intellectual life have receded in apparent importance with the passage of time. Before examining how scientific activities factored into DeWitt Clinton's plans for political advancement, let us revisit the struggles for honor that induced armed combatants to face off atop riverside cliffs.

CHARACTER, VIOLENCE, AND POLITICAL AMBITION

Physically, DeWitt Clinton was at least Thomas Jefferson's equal; he was six feet, three inches tall, with curly brown hair, wide-set dark hazel eyes, a slightly aquiline nose, and a friendly smile. Joseph Priestley's grandson John Finch remarked, "The personal appearance of Governor Clinton is commanding; he possesses more dignity than any other individual I ever saw."[15] Like an Indiana Harrison in the nineteenth century, or a Massachusetts Kennedy in the twentieth, DeWitt Clinton's family connection gave him a gigantic head start in politics. His uncle George Clinton had served as the state's perennial governor, holding that office for twenty-one of the years between 1777 and 1804. As a consequence of his uncle's long and illustrious career, DeWitt Clinton found work in politics before any other vocational choice could divert him. He graduated from King's College at the age of seventeen in 1786, and his uncle hired him three years later to serve as the governor's private secretary. This position, which the nephew filled for the next six years, provided the basis for a meteoric rise.

DeWitt Clinton rapidly ascended the ranks of New York's anti-federalists, winning elections to the state assembly in 1797 and to the state senate in 1798. In February 1802, New York's state legislature chose DeWitt Clinton, aged thirty-two, to join Morris as one of New York's senators in Washington, D.C. Barely a year and a half later, when President Jefferson appointed New York's mayor Edward Livingston to a federal legal post, Clinton relinquished the Senate seat to take

Livingston's place as mayor, a position that afforded more patronage opportunities, as well as judicial and executive responsibilities. Clinton served several more terms in the New York State Senate, simultaneously filling positions of political influence in the city and in the state for much of the time between 1803 and 1815. In 1811, he was even elected to a term as lieutenant governor, a position whose key responsibility was to preside over the New York State Senate.

From the moment he entered the realm of New York politics, Aaron Burr Jr. was DeWitt Clinton's most ambitious rival. Burr was born into a deeply respectable, politically conservative family. He was the son and namesake of the founding president of the College of New Jersey at Princeton, grandson of Rev. Jonathan Edwards on his mother's side, and the first cousin to Yale's staunchly Federalist Rev. Timothy Dwight Jr.—all staunch New England preachers. This birthright was profoundly dislocated when, at the age of three, young Burr lost both parents. A Puritan uncle tried to raise the precociously smart but increasingly rebellious boy. After unsuccessfully running away from home to become a sailor, the child studied intensely and applied for admission to his father's college at the age of eleven. Despite his advanced ability to speak and write in Latin, he was rejected for being physically too small and too young. Unfazed, he pursued the first year of the college curriculum on his own and then won admission to Princeton just two years later, entering as a thirteen-year-old sophomore. Upon his graduation in 1773, Burr briefly considered a career in theology before settling on the law. With the outbreak of war in 1775, he quickly distinguished himself as a fearless and capable soldier in the Revolutionary War.

After the war, Burr rose quickly in civilian society as a lawyer and politician. Having just been elected to the state assembly while New York's ratification of the Constitution was being debated, Burr was appointed by anti-federalist governor George Clinton to serve as New York's first state attorney general in 1789. The next year, again with the governor's support, Burr unseated U.S. senator Philip Schuyler, Federalist father-in-law to both Alexander Hamilton and Stephen Van Rensselaer. Burr served in the Senate for the remaining six years of George Washington's presidency. Returning to the New York political scene in 1788, Burr joined the rapidly advancing DeWitt Clinton as a state legislator. By the next year Burr had wrangled majority control of the assembly on behalf of Jeffersonian democracy, to the dismay of Hamilton and the other New York Federalists. On the strength of these impressive achievements, Burr was selected to serve as Jefferson's running mate in 1800. At this point, however, a reputation for audacious ambition began to alienate Burr from other members of the party.

John Adams hoped, like Washington before him, to be reelected to a second term as president. His aspiration suffered a clear defeat in 1800 with a vote that divided along geographical lines. Populous New York and Virginia joined several other western and southern states to support the Jefferson-Burr ticket. Adams's strongholds were New England, New Jersey, and Delaware. Electors from Pennsylvania, Maryland, and North Carolina split their votes between the two parties. In the final tally, Adams fell just short with only sixty-five of the seventy votes needed to win a majority in the Electoral College. Technically, however, because presidential electors were originally instructed to cast votes for their top two preferences without designating one above the other, both Aaron Burr and Thomas Jefferson were awarded seventy-three electoral votes. The Constitution had provided for unresolved elections to be decided by the House of Representatives. Because some northern Republicans actually preferred Burr over Jefferson, and because it was a lame-duck Congress still dominated by Federalists about to leave office, the House was deadlocked through thirty-five ballots. Jefferson won eight of the sixteen states' votes every time, with Maryland and Vermont abstaining, but nine were needed for a clear majority. Finally, on the thirty-sixth ballot, Hamilton persuaded enough other New York Federalists to shift from supporting native son Burr (whom he personally despised) to Virginia's Jefferson (whom he merely opposed). Burr's reputation began to slide.

Seeking even greater power and influence at the state level, DeWitt Clinton looked for ways to capitalize on Burr's alienation from Jefferson. For example, he used his connections back in New York to undermine the vice president and his closest patronage beneficiary, John Swartwout. On the vice president's recommendation, Jefferson had appointed Swartwout to become a U.S. marshal in 1801. Burr and Swartwout had been close since 1799, when they cofounded the Manhattan Company, a private water company and banking firm that was intended to compete against the Federalists' Bank of New York. Clinton worked with Hamilton and other New York Federalists to quietly gain control over the company. In September 1802, the conspirators arranged for Burr and Swartwout to be replaced as directors at the Bank of the Manhattan Company. When Senator Clinton was named a director in his place, Swartwout guessed who had been instrumental in bringing about his misfortune, and so he decided to provoke Clinton. Usually a master of guile and discretion, Clinton unwisely responded to Swartwout's bait by publicly uttering those deadly words *liar* and *scoundrel*. Clinton's subsequent attempt to apologize was fruitless; the marshal wanted satisfaction. Clinton, who lacked the combat experience so many of his colleagues had brought to their careers in the Senate, had been maneuvered into a duel of honor. On the cliff at Weehawken, three exchanges of pistol fire at a distance of ten yards left both men

unharmed. With the fourth and fifth volleys, Clinton put two bullets in Swartwout's left leg. Despite his wounded opponent's objections that they were not yet
finished, Clinton refused to continue the contest.

Clearly, Clinton had access to finer tools than bullets to outmaneuver Vice
President Burr, including his intellect. After the 1800 election raised perplexing
constitutional and procedural questions, Clinton decided to study the situation
analytically. Exercising his native talent as a logician, he studied the structural
features of the original Constitution, which had permitted the intended election
of a party standard bearer to nearly be derailed by the process wherein electors
voted for both the nominee and his running mate. By December 1803, Senator
Clinton presented the basis for what would become the Twelfth Amendment to
the Constitution; he proposed to separate electoral votes for presidential and vice
presidential candidates. Clinton's measure was passed by both houses and sent to
the states for speedy ratification. Everyone understood that this amendment was
intended specifically to prevent an outcome like Burr's near election in 1801.

Clinton also mobilized his political friends and associates in the press to engineer defeat for Burr and his friends. Early in 1804, though newly returned to New
York City as its mayor, DeWitt Clinton consulted with the president about shifting leadership in New York Republican politics. Contrary to DeWitt's preference
that uncle George remain at the seat of power in Albany, Jefferson invited the elder Clinton to replace Burr as his running mate for the campaign of 1804. The
younger Clinton anticipated that this move might have a paradoxical effect. To be
dumped was to be liberated. Burr might salvage his position as the state's leading
Republican by exchanging places with New York's venerable governor. DeWitt
Clinton worked strenuously to block Burr's chance of taking charge in Albany.
Secretly courting the assistance of New York Federalist leaders including Alexander Hamilton, Clinton persuaded the Federalist Morgan Lewis to relinquish his
seat as New York State's Supreme Court Chief Justice in order to run against Burr
in the spring election. The deal worked. Lewis, not Burr, replaced George Clinton as governor. Within months of this humiliation, Burr killed Hamilton in the
infamous duel that removed both men as factors obstructing Clinton's domina-
tion of New York politics. (Incidentally, the physician who attended to Hamilton's
medical needs at the Weehawken cliffs on that fateful morning was Dr. David
Hosack, Samuel Latham Mitchill's fellow professor of botany at Columbia.)

George Clinton was in his late sixties when he became Jefferson's second vice
president in 1805, and he was no longer altogether healthy. Presiding over the
U.S. Senate, he was neither as vigorous nor effective as Burr had been, but he did
play an active role in exposing Burr's subsequent nefarious activities along the

Texas frontier. Clinton discovered allegations that Burr was scheming, with Mexican assistance, to become the self-styled ruler of a southwestern empire. Jefferson allowed judicial proceedings to go forward in order to determine whether the former vice president was in fact engaging in a treasonable conspiracy to overthrow the U.S. Constitution by violent means. These accusations led to two trials and Burr's eventual disgrace and exile. In the meantime, George Clinton continued to nurture DeWitt's political influence in other ways. For example, whenever the elder Clinton felt isolated from affairs of state, he asked his nephew to serve as an intermediary with the president, which helped retain the younger Clinton among Jefferson's circle. As the end of his second term approached in 1808, Jefferson designated his secretary of state James Madison to be his successor, rather than tapping either James Monroe or George Clinton, who were positioned as potential rivals for the nomination. Despite Clinton's diminishing health, Madison invited him to continue as vice president, an office he held until his death in the spring of 1812.

By that time, Madison was preparing to run for reelection, and the country was on the brink of war with Great Britain. The original two-party configuration had almost completely disintegrated, with Jefferson's party running virtually unopposed at the national level. Belligerent western and southern congressmen like Kentucky's Henry Clay helped to convert Madison's list of grievances into a declaration of war on Great Britain in June 1812. Strongholds of antiwar sentiment existed, especially in rapidly industrializing New England towns and cities, but the Federalists were in no position to mount an effective national campaign. With his uncle's recent death, Burr's disgrace, and the Federalists' apparent disorganization, DeWitt Clinton stepped forward and declared himself as a Republican challenger to Madison. The campaign divided northern Republicans into Clintonian and anti-Clintonian factions. New England Federalists flocked to the New Yorker in the hopes that he would be more likely than Madison to restore peace with Britain. The outcome was remarkably close. Clinton carried his home state, plus New Jersey, Delaware, and all of New England except Vermont. Maryland's electoral vote was divided. Madison enjoyed strong frontier support in the west and south, but he was reelected by an overall margin of just thirty-nine electoral votes. Had the Commonwealth of Pennsylvania tilted north toward New York instead of aligning south with Virginia, the United States would have had its first President Clinton in 1813.

DeWitt Clinton's political ruthlessness and personal recklessness do not in themselves explain why his legacy in American political history is so different from Thomas Jefferson's. He took a major gamble by running for national office against an incumbent during wartime, and in that moment he failed. But there is another side to Clinton's career that supports a comparison with Jefferson's prominence, especially with respect to his role in enabling the investigation of and dissemination of scientific knowledge about the interior of North America. Clinton promoted the importance of natural history to the early national politics and culture of the United States even more effectively than Jefferson had. First of all, Clinton aggressively advocated state-sponsored internal improvements with numerous scientific consequences. (These are explored in chapters 4, 5, and 6.) Beyond that central practical achievement, Clinton also worked hard to cultivate other public and private institutions that would advance science, literature, and the arts.

Clinton's interest in nature was not simply a matter of pretension or public performance. Throughout his life he developed and sustained an extensive and ever-growing private network of scientific correspondents. He advised and educated his own son George to become a proficient naturalist. Most remarkable of all, Clinton personally conducted geological, agricultural, and anthropological observations of New York State that were widely published and respected by contemporary naturalists, demonstrating that he possessed the ability to do the hard work of science. Clinton was thus far more than an office-seeker gauging market forces and political circumstances. He embodied a Renaissance man's vision of civic virtue—one that combined the trappings of power with the exercise of intellectual engagement, literary achievement, and scientific curiosity. From this perspective, his natural history activities are not merely quaint pursuits of historical curiosity but the fulfillment of a coherent ideal of public service. Clinton professed an obligation to seek and sustain novel ways of bringing prosperity and greatness to his constituents. He would find these primarily through the cultivation and exploitation of natural wealth. Even if one doubts Clinton's sincerity, it remains indisputable that he went to extraordinary lengths to secure a handsome reputation as a recognized scholar.[16]

After his failed presidential bid, Clinton refitted his celebrity as a public figure in order to identify and meet the challenges he claimed America faced in repairing its cultural deficiencies. As the mayor of New York City, he had already established himself as an active leader among the city's many learned societies, having organized the Historical Society of New York in 1804, the public school system in 1806, and the Academy of Fine Arts in 1808. As a founding member and the first president of the Literary and Philosophical Society of New York, Clinton directly addressed the state of natural science as a matter of public concern. In the words of his landmark 1814 assessment, he complained that America's "enterprising spirit . . . has exhibited itself in every shape except that of a marked devotion to the interests of science."[17] Samuel Mitchill would later recall this speech, which opened the first volume of the New-York Literary and Philosophical Society's published *Transactions*, as the moment when Clinton burnished his reputation as a leader "proficient in natural and physical science" and not a "mere politician."[18] Clinton actually published similar words more than a decade before his death, puffing forth in the third person about himself: "Mr. Clinton, amidst his other qualifications, is distinguished for a marked devotion to science;—few men have read more, and few men can claim more various and extensive knowledge."[19]

Clinton's support for intellectual and cultural causes in New York was not all that unusual. In some respects, the early republican period was a sort of golden age for civic institutions and public life. It was a time when socially and politically prominent Americans fostered the development of private institutional support for local researches in science. Public-spirited attempts to create literary and philosophical societies, libraries, museums, and lyceums occurred in cities large and small throughout the country. In 1812, for example, Isaiah Thomas, a Revolutionary War veteran, printer, and bookseller, founded the American Antiquarian Society. Due to the impending threat of British invasion, he chose to establish this institution in Worcester rather than Boston. DeWitt Clinton's efforts to found and nurture New York City's Literary and Philosophical Society, the New-York Historical Society, the Lyceum of Natural History, and the Athenaeum were characteristic of what civic leaders in many other cities were also doing.

DeWitt Clinton was not single-handedly responsible for New York's impressive new cultural edifices. Samuel Mitchill, Gouverneur Morris, and another half dozen or more local scientific and political notables were active co-creators involved in the founding and early operation of many of these same institutions. James Renwick, Mitchill's eventual successor as professor of Natural Experimental Philosophy and Chemistry at Columbia, praised Clinton's extraordinary intelligence and foresight: "He is among the few who seem to have seen that the money expended in the support of such institutions is not lost, but will shortly be repaid with interest. In conformity with this enlightened and liberal view, he gave to these societies the benefit of his pen in drawing their charters; his aid as a member of the legislature in procuring the passage of their acts of incorporation; and devoted to their prosperity no inconsiderable share of his time and talent." Renwick added that Clinton's commitment to these learned societies was especially valuable, given his personal reputation: "In these associations, the advantage to be derived from his high political standing, and lofty reputation as a statesman and magistrate, were fully appreciated, in securing unity of action and harmony among persons necessarily rivals."[20]

By spearheading this movement of civic-minded activity in New York, however, Clinton achieved something out of the ordinary; he became widely known as a public figure who both understood and actively participated in scientific research. The effect of Clinton's sponsorship of scientific inquiry and talent was widespread. Caleb Atwater, the Ohio lawyer and amateur archaeologist whose views on the Great Lakes' geological past were introduced in chapter 1, regularly received his encouragement, generosity, and practical scientific advice: "Governor Clinton is procuring me all the books on mineralogy, Geology and organic remains, in addition to those he has, from time to time forwarded me. He advises me to lay aside *all systems and all theories, and to search only for truth.* Perhaps, he anticipates quite too much, from my labours, for he promises me wealth and fame!" (Emphasis in Atwater's manuscript originals was indicated by heavy underlining.)[21] For Atwater, the manner in which politics and science were completely intertwined can be discerned from a private rant he penned to Isaiah Thomas in 1820: "*Mere party politicians . . . promise* every thing and perform *nothing. De Witt Clinton,* is the only, *Honorable* exception within the range of my acquaintance. He loves literature and science, for their own sake, and is equally in love with those who excel in them." Atwater ranked Clinton far above other Republicans, especially the Virginians who had served as president: "I suspect, that some historian will show posterity that Jefferson Madison + Monroe, were deceitful, vicious, and bad men. Monroe is a perfect hypocrite, I fear."[22] This insult seems particularly ungrateful. After all, it was President Monroe, not Governor Clinton, who had appointed Atwater to serve as postmaster for Circleville, Ohio, thus providing a spectacular opportunity for him to partake of regular productive commerce in scientific correspondence and to enjoy the franking privileges that afforded specimen exchanges free of charge!

Atwater expressed more than salutary or honorific praise for a benefactor. He regarded DeWitt Clinton as a powerful political sponsor and a wise scientific mentor. Clinton's practical advice to steer clear of potentially embarrassing theoretical speculation and not to trust any authority other than reproducible truth reverberates throughout Atwater's correspondence: "It requires constant care, I find, to keep clear of errors, sanctioned frequently, by great names. What now, are the theories of Jefferson, Madison + Williamson, concerning our antiquities, but so many beacons, to caution us of breakers on a lee shore?"[23] Atwater's allegiance to Clinton was certainly intertwined with political partisanship, but it is also clear that he was deeply inspired by the quality of Clinton's scientific inquiry, though Clinton's vision of the role science should play in American society was not unusual. Like Clinton, James Madison, Samuel Mitchill, and Hugh Williamson were all Jefferson cronies with reputations for possessing scientifically curious minds. Mitchill and Williamson were intimate supporters of DeWitt Clinton's efforts to cultivate New York's learned societies. In fact, the New-York Literary and Philosophical Society had only come about because Williamson convened a gathering of the state's most prominent men of science. Clinton, Mitchill, David Hosack, steamboat inventor Robert Fulton, and Williamson all shared the desire to give New York an institution comparable to Philadelphia's American Philosophical Society.

For Atwater, no practitioner or patron of science was immune from the danger of systematic self-delusion. Atwater was highly critical, for example, of Mitchill: "I fear our worthy friend Dr. M of New York, is quite too fond of theories. Theories, too often I fear, become beds of Procustes [sic], where facts, are lengthen[ed] or shortened by theorists, to suit them. For my own part, I had rather sit at my ease in the temple of truth, than sail over the whole globe in the balloon of theory."[24] Mitchill had in fact developed quite an embarrassing habit of leavening his public lectures with fantastic theoretical speculations. Frederick Hall, Professor of Mathematics and Natural Philosophy at Middlebury College, complained privately to his Yale counterpart Benjamin Silliman that "Prof. Dr. Senator Mitchill" was a bizarre presenter of science: "The learned Dr., it seems, first takes a skip over the earth, and explains in a thrice, its Cosmogony, its Geognosy, its Mineralogical Chemistry, and its Physical Geography—he dives into volcanoes, and comes out in a flood of scoria, ashes, slag and lava, with a lurid account of their formation." Mitchill's explanations of geology were completely undisciplined, being interlaced with random remarks about "the origin and all the operations of light . . . the ocean . . . how the world will finally take fire, how seas, lakes, rivers, springs, and mineral waters have their origin . . . Aerology . . . mineralogy, and [he] tells us *all about* plants, animals, fishes &c &c." Hall bemoaned how Mitchill's mind arrogantly strayed even beyond the Earth to inventory "*every thing* concerning the sun, and planets, and meteorites & asteroids & comets, and infinite space."[25] DeWitt Clinton avoided this kind of foolishness, preferring to focus upon a very practical natural history that actively linked the intellectual quest for knowledge about the land and its contents to the politics of internal improvements.

On the occasion of his assumption of the presidency of the New-York Literary and Philosophical Society in 1814, Clinton clearly spelled out what his hopes and dreams were in terms of the branches of American natural history: "Men of observation and science ought to be employed to explore our country with a view to its geology, mineralogy, botany, zoology, and agriculture."[26] He spoke with an authority born from personal experience. Clinton had carefully studied research reports published about North America by eighteenth-century European naturalists such as Peter Kalm and Father Hennepin. He understood the principles behind the botanical nomenclature system proposed by Carl Linnaeus. He read and admired the work of Georges Cuvier, and he cultivated a network of scientific correspondence with many other European naturalists, including James Edward Smith, the president of the Linnaean Society of London. David Hosack's eulogy would single out Clinton's knowledge of mineralogy and geology, which he deemed not only highly accurate but superior to that of all but a few persons.

CATALOG OF ACHIEVEMENTS

DeWitt Clinton was not just a well-read amateur. He also had the sort of keen eye that empirical scientists really prized, and the rewards were tangible. When Clinton noticed an unfamiliar variety of wild wheat growing in Oneida County in 1810, he sent a specimen off to the Linnaean Society of London. British scientists confirmed his conjecture that this was a new (i.e., previously unknown) variety of *Triticum*. The discovery earned Clinton honorary membership to that scientific body and provided an entrée into Europe's scientific circles. With a touch of chagrin, Samuel Mitchill noted that Clinton had succeeded in gaining European respect and notoriety far beyond that granted to almost any other American scientist of his day, all for a serendipitous discovery. James Renwick confirmed Mitchill's appraisal, but in a more generous tone, when he recalled: "This discovery of Clinton, therefore, although hardly noticed by his countrymen, procured him much reputation among the learned of Europe; and the diplomas of many societies founded for the cultivation of natural science were showered upon him."[27] Scholarly accolades accumulated. In 1817, David Hosack and DeWitt Clinton were the first Americans honored with memberships to Great Britain's Royal Horticultural Society.[28]

Like Jefferson's investigations into North American antiquities, Clinton's scientific discovery contributed to American national pride in a symbolically powerful way. The wild wheat discovery created an intellectual sensation because *Triticum* had both botanical and anthropological significance. By identifying an indigenous species of *Triticum*, Clinton challenged contemporary theories about the place of origin of human beings on Earth. Western Asia had been regarded as the cradle of the human race, and the prevalence of belief in this view was based on the simple observation that wild wheat plants are native to the shores of the Caspian Sea. If Oneida County's wheat was truly unrelated to Old World varieties, then North America might independently have nurtured the birth of its own civilizations based upon agriculture. In other words, North American antiquities could be seen as products of a completely independent culture situated in a landscape whose natural origins owed nothing whatsoever to the Old World. Contemporaries inferred with patriotic pride that if Oneida's wheat was both spontaneous and indigenous, New York's claim to fame would rival that of Sicily, birthplace of the mythological queen of the harvest, Ceres.

The wheat discovery was no fluke. Clinton investigated the practical agricultural potential of many other native species. For example, he identified a species of wild rice growing in the Montezuma swamps of the Seneca River as *zizania aquatica*. *Zizania* was known to flourish in many North American lakes. "Some of the Western Indians derive their principal support from it. The grain it bears is superior to the common rice, and if cut before ripe, it makes excellent fodder, embracing the advantages of hay and oats."[29] Clinton concluded that North America's botany provided a "vast ability to support the human species, and of the propriety of calling its latent powers into operation."[30] In other words, Americans had access to a bounty of New World organisms that, if scientifically assessed and utilized, might well establish an independent basis for prosperity not beholden to European imports. Continuing along these lines, Clinton produced original reports of newly discovered species of fish (the *Salmo Otsego*, for example) and birds (a swallow designated as *Hirundo Fulva*, which John James Audubon later identified independently).

Clinton's skill in making such discoveries and his persistence in publishing learned accounts of them helped the politician garner the respect of other practicing scientists who recognized the importance of having not only a keen eye but also the scholarly discipline required to distinguish something new in the field from the compendium of things already known. In sum, Clinton cared deeply about New York's natural history, and he knew how to engage in field research according to the standards of his era. He repeatedly demonstrated his knack for recognizing the key piece of physical evidence in the landscape and his alertness to its scientific as well as its practical implications. Moreover, Clinton was deeply concerned about the extirpation, if not the extinction, of species native to his state. He understood that populations of freshwater fish were an important component of his state's natural wealth, no less so than the mineral and agricultural productions to be wrested by human ingenuity from its terrain.

The breakthrough moment for Clinton's transformation into an accomplished natural historian occurred in 1810. His "Private Canal Journal" from that summer contains extended ruminations about New York's geography and stratigraphy. Though sometimes flowery in their speculations, his descriptions of specific locales consistently met the scholarly standards of contemporary geological reports. Clinton's original analysis of geology and topography at Little Falls brought together specific, technical, eyewitness descriptions of mineralogy, literary classical allusions, a metaphorically rich and dynamic syntax, and a providential tone hinting at a Creator's benevolent purpose. This version of the "lakes bursting through mountains" hypothesis was what Darby found so compelling.

> As you approach the falls, the [Mohawk] river becomes narrow and deep, and
> you pass through immense rocks, principally of granite, interspersed with lime-
> stone. In various places you observe profound excavations in the rocks, worn
> by the agitation of pebbles in the fissures, and in some places, the river is not
> more than twenty yards wide.
>
> As you approach the western extremity of the hills, you will find them about
> half-a-mile from top to top, and at least, three hundred feet high. The rocks
> are composed of solid granite, and many of them are thirty or forty feet thick,
> and the whole mountain extends, at least, half-a-mile from east to west. You
> see them piled on each other, like Ossa on Pelion; and in other places, huge
> fragments scattered about in different directions, indicating evidently a violent
> rupture of the waters through this place, as if they had been formerly dammed
> up, and had forced a passage through all intervening obstacles. In all directions
> you behold great rocks exhibiting rotundities, points, and cavities, as if worn by
> the violence of the waves or pushed from their former positions.
>
> The general appearance of the Little Falls indicates the existence of a great
> lake above, connected with the Oneida Lake, and as the waters burst a passage
> here and receded, the flats above formed and composed several thousand acres
> of the richest lands.[31]

Echoing Jefferson's geological speculations about a violent rupture of the Shenandoah's waters through the ridges at Harpers Ferry, Clinton's rhetoric displays here a considerably more advanced grasp of the modes of contemporary geological discourse. By taking the extra step of thinking beyond his geological observations to explore the practical consequences of understanding Earth's history, Clinton implied that the location of fertile lands would have much to do with the boundaries of these former lakes. At the same time, it is clear that Clinton treasured the intellectual challenge of spinning plausible tales of geological conjecture. His description ended wistfully: "This great Lake—breaking down in the first place to the east, the place where its waters pressed the most, and then to the west, where its recession was gradual—forms an object worthy of more inquiry than I had time or talent to afford."[32]

Clinton exercised scientific habits of mind even when he was heavily burdened with affairs of government. He continued to make reports to learned journals about his own scientific opinions and discoveries regardless of which executive and legislative offices he occupied. Like Jefferson, Clinton's bid to embrace leadership positions in science and politics did not escape challenge. When he became governor in 1817, Clinton's enemies mounted a campaign to embarrass him about the incongruity of simultaneously holding political and intellectual leadership posts. At that moment he was the president of the New-York Literary and Philosophical Society and the Academy of Fine Arts. To complicate matters, the New York Historical Society, of which he had been a founding member in 1804, had just selected him to succeed the recently deceased Morris. The first intrusion of partisan politics upon learned society politics occurred when New York's brand new Lyceum of Natural History, founded under the presidency of Clinton's friend Mitchill, suddenly rejected Clinton's candidacy for an honorary membership. Clinton had advocated that all these organizations receive public support, and he did not intend, as a sitting governor who would be the target of many enemies, to compromise them. Stung and surprised by the Lyceum rebuke, he chose to resign all his presidencies except that of the Literary and Philosophical Society. The Academy of Fine Arts presidency went to the history painter John Trumbull and the Historical Society presidency went to David Hosack. Neither organization continued to thrive so well as they had under Clinton.

Clinton realized that he had to be more careful when publishing his scientific notices, especially when they doubled as politically charged essays. In 1820 he composed a series of letters using the pseudonym Hibernicus. These letters, first published in serial form in the *New York Statesman*, demonstrate his remarkably wide-ranging scientific intellect. He succinctly framed an open question then enlivening contemporary geological discourse, for example, when he wrote: "Whether these interior [Finger] lakes have been formed from the retreat of the ocean, and are in a state of gradual subsidence; or whether they have been produced by springs and deposits of water in great cavities, enlarging gradually their dimensions by breaking down the feeble barriers of *schist* with which they are surrounded, are still points *sub judice*." In another passage, Clinton introduced the ongoing debate about Niagara's erosion: "The recession of the falls from Lewiston and Queenston, is easily explained on this geological view of the country. The fragile materials which compose the foundations of the great calcareous rocks are continually and gradually wearing away by the action of water, and by a partial exposure to the atmosphere; the removal of the sub-strata will necessarily produce a precipitation of the super-incumbent rocks into the watery gulf." But he did not settle for this simple answer. With surprising foresight, he suggested that the agency of ice might be implicated: "The progress of this operation is obvious—the immense bodies of ice which are carried down from Lake Erie, must also be a powerful auxiliary, and frost and earthquakes unquestionably contribute greatly to the production of these results."[33] The authorship of the Hibernicus letters was privately known by many, including Mitchill, who publicly commended their treatment of "miscellaneous matters in zoology, mineralogy, geology, and the kindred sciences; proving the author to be a careful observer, who registered in his journal, not only the facts and occurrences before his eyes, but also the analogous appearance in other parts of the globe."[34]

Among celebrations of American land, its mineral contents, and speculative answers to the mysteries of Earth's history, Clinton embedded a geological rationale for constant terrestrial change: "I find the geology of this country most extraordinary; it is *sui generis*. In using the technology of Werner, I beg you to understand that I am no disciple of his school. I adopt it to explain my ideas in conformity to received and general nomenclature. We are yet in the horn book of this science. The lapse of ages will accumulate facts for the formation of systems.—This earth is undoubtedly a wreck of a former world; a new combination of old materials. Fire and water have been the principal agents in accomplishing this work; and changes are constantly going on, sometimes with slow, at other times with rapid, and always with unceasing steps."[35] Dynamic historical geology had great ideological potential. In the views put forth by Hibernicus, the North American landscape not only testified to its own uniqueness but also hinted at natural analogues for both revolutionary and gradual social change. Americans had successfully waged a war of independence, but they had thus far imperfectly mastered republican government. Just as the rocks had seen suddenly violent and imperceptibly cumulative changes, so would the politics of this young country. Clinton had been born too late to partake of the initial melee against the mother country, and he had stood responsibly but diffidently by when hostilities were renewed in 1812. Now, the work of the next generation of American leaders was clear, and it was all about fostering intellect, ingenuity, and internal improvements.

Hibernicus urged his readers to look to themselves and their own land for answers to the mysteries of existence. Americans needed to stop looking across the Atlantic for clues about how to proceed. "If America will not stand on its own legs, and rely on its own exertions, what can it expect but supercilious arrogance

and contumelious assumption?" Hibernicus predicted that independence in science would follow independence in politics: "Dr. Silliman's periodical work on
Natural Science is superior to any thing of the kind published in Europe; and
there are men of genius and of learning in every section of the country, who with
an adequate encouragement would redeem the American character from the
obloquy of transatlantic insolence."[36]

Herein, however, lies a paradox. Clinton clearly craved and cherished his international reputation as a man of science. Europe remained the inescapable
standard for all cultural achievements. Embracing the Latin designation for
Ireland, Clinton's persona was implicitly a distinguished visitor from the British
Isles, a man whose great knowledge of the world and positive impressions would
bring prestige to New York and shower foreign respectability upon the state's local
pride. In clothing himself as Hibernicus, Clinton reinforced the notion that only
a foreigner could persuasively pronounce upon the quality of all things American,
like *Silliman's Journal*, as compared with their European counterparts. At the
same time, Clinton actively deployed natural history investigation as an important tool to advance New York's comparative development among the United
States and as an expression of national patriotism. Support for this claim can be
found in historian Charlotte Porter's passing observation: "Clinton complained
that American species of birds 'were lost under European names.' He advocated
statewide research of plants, animals, and mineral resources, studies [made] practical for the first time along the new canal excavated across the state during his
gubernatorial administration."[37]

These behaviors fit somewhat uncomfortably within the idealistic Enlightenment worldview DeWitt Clinton must have imbibed from his early education.
At the start, he had placed tremendous faith in the agency of human ingenuity over Nature. As early as 1794, Clinton professed a powerful vision of how
the international fraternity of republics would work together to ultimately transform continents: "Great improvements must also take place which far surpass
the momentum of power that a single nation can produce, but will with facility
proceed from their united strength. The hand of art will change the face of the
universe. Mountains, deserts, and oceans will feel its mighty force. It will not
then be debated whether hills shall be prostrated, but whether the Alps and the
Andes shall be levelled; nor whether sterile fields shall be fertilized, but whether
the deserts of Africa shall feel the power of cultivation; nor whether rivers shall be
joined, but whether the Caspian shall see the Mediterranean, and the waves of
the Pacific lave the Atlantic."[38] The unpredictable manifestations of revolutionary
France's violence soon diminished this youthful sense of giddiness, but if New

York was ever to enjoy the fulfillment of a destiny that Clinton saw written into its very rocks and plants, its citizens would have to learn to live up to their divinely ordained promise. As the anonymous Hibernicus, Clinton could trumpet the potential for American science, crediting it with a greatness that was, from an impartial perspective, far from proven. As New York City's mayor and especially as the state's governor, he would consistently seek and exploit opportunities to provide substantial support and inspiration to natural historians in order to salvage the magnificent charge his generation had inherited: to bring civilization to a greater state of fulfillment on the blessed land that it had been given in the New World.

JOHN FINCH'S VISIT TO ALBANY

The recipe Clinton adopted to advance his ambitions may appear to be so sensible that no twenty-first century reader would give it a second thought. He hoped to wed technology and material prosperity to politics. From a modern perspective, it is not surprising that scientific knowledge was significant in Clinton's ascent to power. Upon closer inspection, however, the quest for detailed geological knowledge and the enormous personal and political investments Clinton made in cultivating his own credibility as a patron and practitioner of natural history can seem extreme. Standard historical narratives about internal improvement in the early American republic, voluminous and perceptive as they have been, have not adequately explained the conditions that brought together plans to promote economic and social prosperity, partisan politics, and the rapidly changing science of geology. A fresh starting point for this inquiry can be found in the travel writings of John Finch, an English amateur geologist who chose to visit North America in 1824–1825. "Previous to leaving the shores of England, when I thought of my visit to America, I was anxious to see and converse with De Witt Clinton, whose fame had travelled across the Atlantic to Europe. A friend at New York, to whom I had letters, politely offered to introduce me, and I eagerly availed myself of the opportunity. I wished to see the individual to whom America owed more than any one not engaged in the War of Independence. The canals in every part of the United States owe their successful commencement to his talents and persevering industry."[39]

John Finch had journeyed all the way from London to see natural wonders and prominent men of science in the Empire state. He was not disappointed: "The personal appearance of Governor Clinton is commanding; he possesses more dignity than any other individual I ever saw." Finch happened also to be the grandson of Joseph Priestley, the world-renowned English chemist and Unitarian heretic who had spent his final years as a refugee natural philosopher residing in the remote town of Northumberland, Pennsylvania. Now Finch was making his own bid for fame and honor in the natural sciences. As if to reassure any American readers, Finch's account reverently explains why he expected to encounter a philosopher-king in Albany: "Governor Clinton is known in Europe by his literary productions, and by the part which he has taken in the New York canals. Since the earliest period of active life, in the midst of numerous avocations, he has favored science and literature by every means in his power."[40]

Almost before it had begun, Finch's private interview with Governor Clinton transpired into a colloquy with yet another New York politician. Clinton offered to introduce his visitor to Stephen Van Rensselaer III.[41] Van Rensselaer, Albany's representative to the U.S. Congress, also happened to be the wealthiest private landowner in the United States. Known as the Patroon, he had inherited his title from a Dutch ancestor, and his vast estate started as a colonial land grant along the Hudson River in the seventeenth century. Congressman Van Rensselaer would very shortly "decide" the controversial presidential election of 1824, earning himself a footnote in all the history books for a reason having nothing whatsoever to do with his vast riches.[42] In the glow of a lovely autumn morning, Finch and Clinton strolled along the streets of old Albany northward to meet the old patrician Federalist at Rensselaerwyck Manor. Along the way, Finch peppered Clinton with questions designed in part to flatter the statesman, but Finch was also exploring and testing his own notions about the natural circumstances of empire. When Finch asked Clinton what he had learned about comparative military advantages from the war recently fought between New York militia and British regular troops, Clinton noted only that invaders "who had been a long time cooped up in ships" might be physically weaker than freshly mobilized defenders. When asked to predict what city would benefit most by the canal, the former governor lit up with pride: "Imagine a country equal to Great Britain and France in extent, exporting all their superfluous produce, and importing merchandise and foreign supplies by one port. That port is New York! She will be the greatest commercial city in North America."[43] Readers back in England could smirk at the self-importance of Clinton's vision for the former colonial seaport; little did they suspect that New York's dreams of wealth and grandeur would, within mere decades, rival those of London itself.

The manor house had been built "country style" upon the flats east of the river road near where Patroon's Creek flowed into the Hudson. Approaching from the south, the visitors would first have glimpsed the mansion's hipped, English-derived gambrel roof with pedimented dormers. These were perched atop an expanse of solid masonry, with octagonal wings recently attached to either side; these, along with the outbuildings, symmetrically framed the main entrance.[44] The Patroon escorted his two guests into the great hallway. Finch was astonished to learn that Van Rensselaer's estate now comprised some two hundred square miles of towns, fields, and forests, upon which some three counties' worth of tenant farmers lived by virtue of a basically feudal quit rent arrangement. But their landlord was no relic of the medieval past. Van Rensselaer was an astute and voracious reader, remarkably conversant with the recent doings of Europe's leading scientific thinkers. He had graduated from Harvard College in the class of 1782, served in a variety of public offices throughout his adult life, and was the major philanthropist responsible for funding many Albany and Troy cultural and civic organizations.

Most impressive of all to Finch was Van Rensselaer's track record as a patron of science. The expenditure of some £1,000 sterling to finance a series of ongoing geological surveys across the state proved that the Patroon's private investments compared with Clinton's public efforts. Finch could not have been more delighted had he been transported to the Salomon's House of Sir Francis Bacon's utopian imagination. Although Finch does not invoke Bacon in his account, virtually every American scientist of the early nineteenth century paid homage to the progenitor of a mechanistic and inductive program of practically valuable empirical science. As the feminist historian of science Caroline Merchant explains, "The Baconian program, so important to the rise of Western science, contained within it a set of attitudes about nature and the scientist that reinforced the tendencies toward growth and progress inherent in early capitalism."[45] New York's leaders had presented themselves as builders of an enlightened society of prosperity, based on the practical benefits to be derived from the systematic scientific investigation of nature.

What enabled these two elite gentlemen of upstate New York to contend for such honors, privileges, and influence, and how was the practice of natural history implicated in their quest for power? Answers to these questions lie in the history of ideas and practices in the earth sciences prosecuted under their auspices. To analyze the "geological imperative" requires interactive historical reasoning, a process that examines both circumstances permitting and consequences following those scientific ideas and practices. DeWitt Clinton and Stephen Van Rensselaer came from distinct ideological, ethnic, educational, and class backgrounds and therefore disagreed about many things. Nevertheless, these two leaders shared a vision of New York's natural potential to yield society's three most highly prized commodities: prosperity, political power, and prestige. They agreed that knowledge, specifically scientific knowledge, was the key to prosperity.

Prosperity in the early republic was widely understood to involve more than just money. As well as material comforts, the well-to-do person enjoyed such things as a fine reputation among one's peers, a large family, and the time and resources needed to nurture one's own intellectual and spiritual growth. Clinton and Van Rensselaer also used their power and prestige, both individually and in concert, to nurture change. Because these two privileged men acted as patrons of technological and institutional innovation, paying their highest respect to natural knowledge, Albany became an epicenter for investigating, teaching, and applying the lessons of natural history. As a consequence, New York sustained its original investments in the development, assertion, and deployment of geological knowledge, commitments that in turn spawned an influential set of assumptions about how science and prosperity can and should be linked together.

CHAPTER THREE

The Landlord and the Ex-convict

Who will remember the beneficent Stephen Van Rensselaer? The myriads of ambitious mortals who have preceeded [sic] us are forgotten. So we of the present generation who are wearing down our strength in climbing precipices and descending caverns, cannot hope to be remembered but a few years. Why should Van Rensselaer send us here at great expense, when he too is so soon to be forgotten? It must be that he has ungovernable propensities to do good, which are as unmanageable as the thirst of the drunkard.

—*Amos Eaton, private journal, 1824*

Stephen Van Rensselaer is largely forgotten today, but he was an extraordinarily rich and powerful man who influenced pivotal events in the early American republic. Born in New York City in 1764, Van Rensselaer was brought up among the colonial aristocracy. He was initially sent, as Aaron Burr had been nine years ahead of him, to Princeton College but completed his studies at Harvard in 1782. Young Stephen's prominence in society was guaranteed. As scion of the Patroon of Rensselaerwyck, Van Rensselaer inherited the largest landed estate in North America, a remnant of the Dutch system of colonial land grants. The new Patroon would become America's first millionaire, but unlike men who got rich (fairly or unfairly) on land speculation in the unsettled West, Van Rensselaer's land empire was concentrated in the midst of New York State's capital district. Over the course of his life, he accumulated and maintained the legal title to virtually all the private property in New York's Albany and Rensselaer counties, as well as half of Columbia County. Owning everything in these long-settled and populated counties, Van Rensselaer bore a closer resemblance to a medieval feudal lord than anyone else in the history of the United States. One contemporary visitor from Poland made the comparison explicitly: "The word *Patroun* is pronounced by everyone with deference and a certain fear. He holds here the place held by the Radziwills in Lithuania."[1]

NATURAL ARISTOCRAT

Like other Hudson valley manorial landowners, such as the Livingstons and the Van Cortlandts, Van Rensselaer pursued a career in politics—a textbook case of *noblesse oblige*. Though he was a Federalist, his political fortunes were curiously intertwined with those of the anti-federalist Clintons. Van Rensselaer launched his political career in 1789, serving as a twenty-five-year-old elected representative to the New York State Assembly. DeWitt Clinton would initiate his political career in much the same manner eight years later, when he was twenty-seven. Meanwhile, DeWitt's uncle George was firmly entrenched as New York's first state governor. In 1794, the nation's chief justice stepped down from the Supreme Court to challenge George Clinton. New Yorkers responded positively, electing as their first team of Federalist chief executives Governor John Jay and Lieutenant Governor Stephen Van Rensselaer. When Governor Jay decided to retire in 1801, the party chose Van Rensselaer to run, but he was defeated by a resurgent George Clinton. The Patroon's extensive domain turned out to be a mixed blessing in this election, for while it ensured influence over his many tenants, it also suggested that he was out of step with the country that had just swept advocates of Jeffersonian democracy into power in many states.

A decade later, the War of 1812 dramatically altered Van Rensselaer's leadership prospects. The New York governorship had changed hands and parties with some frequency during the first decade of the nineteenth century, consistently favoring the Clintons' interest over Van Rensselaer's. DeWitt Clinton's support had enabled the Federalist jurist Morgan Lewis to defeat Aaron Burr in 1805. Governor Lewis was replaced in 1807, however, by Clinton's newest Republican protégé Daniel Tompkins. Tompkins was reelected in 1809 and again in 1811, bringing DeWitt Clinton along with him this time to serve as lieutenant governor. In June 1812, war was declared. Like most Federalists, Van Rensselaer opposed the Republican rush to fight a war with Great Britain, but he was nevertheless immediately caught in the snare of patriotic duty. Historians surmise that Tompkins selected Van Rensselaer to serve as major general of the state's volunteer militia because it would be an opportunity to embarrass the Federalist most likely to challenge his bid for reelection the following year. Alan Taylor likens the situation to "a political game of chicken" in which Van Rensselaer had no good choice: declining the offer would lead voters to doubt the Patroon's patriotism; accepting would mute his criticism of the war.[2] Historian Evan Cornog characterizes the quandary

presented by the Republican governor in an even more unflattering light: "Tompkins appointed the patroon to command the state's troops at the outset of the war,
when preparation was so poor that Napoleon himself would have been hard put
to organize a successful march, let alone fight a battle."[3]

Whether or not Tompkins intended it, this opportunity to acquire battlefield
acclaim did in fact injure Van Rensselaer's reputation. President Madison was
desperate for a quick American military victory on the Canadian frontier, especially after William Hull's ignominious surrender at Detroit in August 1812.
Having been awarded command of the Army of the Center, Van Rensselaer
might reasonably have suspected that his fellow New Yorkers were setting him
up for failure. Peter Buell Porter, Republican congressman from Buffalo, acted
incompetently as the assistant quartermaster general, and the troops were poorly
equipped and unprepared for battle. General Van Rensselaer complained bitterly
on behalf of his shoeless soldiers: "Our best troops are raw, many of them dejected
by the distress their families suffer by their absence, and many have not the necessary clothing. We are in a cold country, the season is far advanced . . . With my
present force it would be rash to attempt an offensive operation."[4] Overriding all
these concerns, Madison insisted that an invasion of British Canada be launched
across the Niagara River.

Recognizing his own very limited experience as a military field officer, the
Patroon appointed his cousin Solomon Van Rensselaer to lead the assault force
of three hundred New Yorkers. Solomon was a veteran soldier who had recently
seen action with General William Henry Harrison's company at the 1811 Battle
of Fallen Timbers. On 13 October 1812, Stephen Van Rensselaer dispatched this
vanguard across the treacherous river despite heavy enemy fire. His gallant cousin
sustained five leg wounds while leading the troops successfully up the several-
hundred-foot embankment on the Canadian side of the river, those same Queenston Heights whose geological history Peter Kalm and others had speculated
about for decades. American soldiers attained the summit, but the battle turned
after British and confederated Indian reinforcements arrived from Fort George. A
contingent of New York militiamen still waiting to cross the river decided abruptly
that they could not be compelled to fight on foreign soil. Regardless of the question of whether his men were legally obliged only to defend the state, their
inaction demonstrated that Stephen Van Rensselaer lacked the charisma and
authority to effectively lead the invasion. Ironically, the first wave of American
invaders had managed to kill the opposing British general Isaac Brock, but now
they were trapped atop the opposing shore's promontory by enemy forces. Some
of the stranded Americans leaped off the cliff into the turbulent river. Those who

did not escape or die were forced to surrender. The Patroon resigned his commission shortly after this humiliating defeat and in due course lost his second bid to become New York's governor in 1813.

Peacetime provided far better opportunities. Despite their political differences, Van Rensselaer shared DeWitt Clinton's sincere passion for the agricultural and geological sciences. Unlike Clinton, the Patroon commanded a private fortune that allowed him to sponsor scientific research projects on a scale that the state legislature was as yet unwilling to fund publicly. Van Rensselaer's commitment to the acquisition and dissemination of natural history knowledge make sense, given his vested interest in land. His private scientific patronage imitated the support that wealthy European aristocrats had traditionally provided natural philosophers. In the end, however, his most profound influence on the intellectual life of the United States lay in establishing a new kind of institution for higher education—one devoted to practical scientific training and the cultivation of engineers. The founding genius, Rensselaer's partner in creating the first civilian "polytechnic" college in the United States, was a hapless lawyer named Amos Eaton.

BOTANIST BEHIND BARS

Born just forty-eight days before the Declaration of Independence was signed in 1776, Amos Eaton belonged to the first generation of citizens who would live their entire lives in an independent United States. American social historian Joyce Appleby has characterized these Revolutionary War babies as a cohort saddled with an identity crisis.[5] Eaton grew up roaming the foothills of the Taconic mountain range near his birthplace in Chatham, New York, before attending Williams College in Massachusetts. Graduating in 1799, Eaton was just a couple of years older than classmate Caleb Atwater, who would later venture out to Ohio to make his mark in law, frontier diplomacy, and science. Like Atwater, Eaton considered entering the clergy upon graduation, but the fervor of the Second Great Awakening, the Protestant religious movement that swept through the region during the 1790s, left Eaton disaffected. Instead of becoming a minister, he planned to get married right away and needed a secure income to be able to provide for his family.

As a young adult, Eaton's private affairs unfolded with a complexity worthy of an eighteenth-century novel. Mistakenly believing that his childhood sweetheart Sally Cady was betrothed to his best friend Elijah Thomas, Eaton wed Thomas's sister Polly instead. The couple moved to New York City where Eaton began legal studies with Josiah Ogden Hoffman, a close associate of Alexander Hamilton.

Shortly after Eaton was admitted to the bar in 1802, Polly, who had always been sickly, died of pulmonary consumption (tuberculosis). The young widower and his three-year-old son Thomas moved upstate to Schoharie County, where Eaton set up a small law practice and land agency. He then asked his first love, Sally Cady, to become his wife, and the family relocated to the beautiful little town of Catskill in 1804.

Domestic bliss prevailed for the remainder of that decade, with four more Eaton boys arriving at the rate of one nearly every other year. Professionally, however, Eaton's life was not so smooth. Inept in his business dealings, he managed to antagonize a gang of real estate speculators who had ties to Nathaniel Pendleton, a powerful New York judge. On 7 January 1811, Eaton was indicted for the crime of forgery and arrested. His accusers alleged that he had misdated a 29 November 1806 land sale document so that it would appear to have been signed on 1 June 1811, a day that had not yet occurred! By the time his case reached trial in May, the prosecution had manufactured the evidence it needed. A Catskill jury convicted Eaton of forgery in conjunction with the now foreclosed property. By 26 August, the forlorn thirty-five-year-old found himself locked up in New York State's Newgate Prison at the head of Tenth Street in Greenwich Village, contemplating his harsh sentence of punishment "at hard labor for life."

The first year and a half of his sentence brought Eaton unremitting despair. First, his mother died while he awaited trial, and then his father died in October 1812. He was unable to attend either funeral. The routine at New York's first penitentiary, considered by contemporaries to be modern and enlightened in its design, was brutal. According to a memoir later published by a fellow Newgate convict, from the moment of arrival, one's life became all but forfeit. "[H]e is immediately put to work, and kept at hard labour, agreeable to his sentence. In summer the rooms are unlocked at 6 o'clock in the morning; in winter at day-light, when the prisoners are called to work, at which they continue till 6 o'clock in the evening, allowing sufficient time for their meals, which are three every day. On the beat of a drum, at 9 o'clock in the summer, and 8 o'clock in the winter, they retire to bed."[6] There were precious few of Eaton's social and intellectual caliber with whom to commiserate, for Newgate was intended for felons only. A person could be imprisoned for life for any of the following crimes: "Rape, robbery, burglary, sodomy, maiming, breaking into and stealing from a dwelling house, some person therein being put to fear, forging the proof of a deed, or the certificate of its being recorded, forging public securities, counterfeiting gold or silver coins."[7] Adult males, female criminals, juvenile delinquents, and the criminally insane were all housed eight to a room (twelve by eighteen feet).[8] Men and women were, of course, segregated. Solitary confinement was an expensive punishment that the wardens reserved for only the most refractory and dangerous prisoners.

Late in 1813, while the rest of the country was caught up in the war with Great Britain, Eaton underwent a spiritual rebirth. Although he later attributed his renewed hope to the ministry of the prison's chaplain, Mr. Milldoller, Eaton's redemption began when he resolved to commit himself fully to the study of nature, regardless of the difficulties imposed by his confinement. He had always been a gifted observer of nature, devoting many hours to wandering the hillsides along the Hudson River. From the fruits of these rambles, Eaton began to assemble a collection of notes that he hoped to publish as an amateur manual of botany. He had even tried to open a Botanical School at Catskill back in 1810, but this plan collapsed as soon as his professional legal reputation was attacked.

While Eaton languished in custody awaiting trial in the spring of 1811, Sally gave birth to yet another son, Eaton's sixth. All of his offspring were named in honor of people he respected, admired, or loved, a practice that provides strong clues regarding his shifting values system over the course of a lifetime. His fourth son, born in 1809, was named Timothy Dwight Eaton after the Yale luminary. From prison, Eaton sent instructions that this sixth boy be christened Charles Linnaeus Eaton, in honor of the great eighteenth-century Swedish botanist Carl Linnaeus, Peter Kalm's teacher, who had promulgated the binomial Latin nomenclature system intended for the classification of all living organisms. Sadly, Eaton would never get to see his baby Charles, who died before reaching the age of three.

Eaton began his study by devouring the few scientific resources available to him through the prison's modest library and building relationships with any and every person who could assist him in his quest for knowledge. Eaton detailed the changes to his regime in a letter to his wife in February 1814. After each long day's labor in the prison workhouse, Eaton stayed awake for hours to pursue his science by candlelight: "I spend my evenings in progressing with my Botanical work, which I commenced in Catskill. I have contrived a new method of arrangement, by which I can exhibit all the known species of plants (about forty thousand) in one small duodecimal volume. So that you can readily determine the name of every plant and its uses in medicine, diet, agriculture and the arts by merely inspecting the flower and some few other parts." Though he soon exhausted the relevant holdings of the prison library, Eaton's interest in botany attracted the notice of the prison agent's seventeen-year-old son, John Torrey, who offered to supply the convict with books he might want to consult.[9] Eaton was confident about the practical utility of his approach to science, a trait that would sometimes

verge on audacity, but which also gave him the freedom to pose fruitful hypotheses. Torrey would eventually found the New York Botanical Garden, become
the leading American botanist of his generation, and remain Eaton's lifelong
devoted scientific friend and confidant.

Having scoured the prison library for works on natural philosophy (incred-
ibly, it possessed a copy of the German laboratory chemist Friedrich Accum's
1802 *System of Theoretical and Practical Chemistry*), Eaton requested that Torrey
obtain an edition of Irish geologist Richard Kirwan's 1794 *Elements of Mineralogy*. This relatively scarce work had special meaning for Eaton, for he had first
become acquainted with Wernerian ideas about geology, which attributed the formation of all rock types to the action of water (sedimentation and precipitation),
by transcribing a borrowed copy of Kirwan in 1802, that painful year when his
young first wife Polly Thomas had died.[10] Perhaps seeking to recapture a familiar
feeling of therapeutic distraction, Eaton now immersed himself in compiling
and cross-referencing all the information he could glean from these authoritative
European works on natural history. From 6 December to 21 December 1814, and
again from 16 January to 6 February 1815, Eaton worked feverishly to produce a
346-page manuscript. Building upon his annotations of the Kirwan and Accum
texts, he appended four pages of notes on Werner's *Mineralogy* to his compilation,
which he entitled "A System of Mineralogy."[11] This handwritten notebook, which
is preserved among the Eaton papers at the New York State Library, was meticu-
lously organized. Eaton evidently intended it to serve as a practical guide for how
to write up one's field notes. Did he imagine that he would ever get to use the
system himself, or were these just the pathetic labors of an eccentric, hopelessly
incarcerated person?

Though he could not have anticipated it, Eaton's intellectual companionship
with Torrey rekindled the possibility of a happier life. John's father William Torrey
was not only the warden at Newgate Prison but also an alderman. At that time,
New York City's mayoral duties included intimate oversight of the judicial system.
When the elder Torrey mentioned that his son had befriended a studious, scientifically minded prisoner, Mayor Clinton was intrigued. Gaining an interview
with the powerful politician, a man already enamored of natural history and heavily committed to the cultivation of learned societies, was an enormous stroke of
good fortune for Eaton. Under normal circumstances, these two men might have
experienced several obstacles to the formation of any kind of personal relationship: Eaton had come from a traditional stock of New England Congregationalist
farmers and soldiers, a demographic that tended heavily toward Federalist and
away from Republican political sentiments. Despite their differences, however,

both men shared a passionate interest in all aspects of natural history. Though no record seems to have survived of their first conversation, it is not hard to imagine that the convict made the most of his unusual opportunity to win DeWitt Clinton's respect, sympathy, and support.

REBIRTH AS AN EDUCATOR AND FIELD RESEARCHER

In 1815, diplomatic and military events thousands of miles away conspired with shifting political fortunes closer to home to make Eaton's forlorn dream of a life in science suddenly viable. On Christmas Eve, 1814, a peace treaty with Great Britain was finally secured in Ghent, Belgium, halting two and half years of war and lifting the clamp from more than a decade of depressed Atlantic trade. Two weeks later, Andrew Jackson's troops won a decisive, after-the-fact victory over British regulars at New Orleans. Upon learning of these two happy events and the subsequent ratification of the treaty by Congress in mid-February, an ebullient New York public welcomed a renewal of peacetime ways and the promise of prosperity. With the public mood now well disposed to amnesty, Governor Tompkins issued a blanket pardon on 17 November 1815 to all nonviolent felons incarcerated at the state penitentiary, including Amos Eaton. The only condition attached to Eaton's release was the stipulation that within three months "he depart from the state of New York and never thereafter return to the same."[12]

At the time of his liberation, Eaton was thirty-nine. He had been granted a second chance to make good on the scientific career he had begun despite such dim prospects, and it was now his ambition to become one of the most respected and skillful practical naturalists in the United States. In attempting to fulfill this dream, he would also seek to establish his own system of geological thought and practice, efforts that yielded, among other fruits, the first geological map of New York State. This does not mean that the remainder of his life was easy. The strain caused by his imprisonment and the death of their baby had rendered his beloved wife Sally dangerously melancholy. Within a few months after their exile from New York, she was dead.

Eaton went first to Connecticut to obtain the only formal training in science he would ever receive, spending one term at Yale College studying mineralogy and chemistry with Benjamin Silliman and burnishing his native gift for botany under the approving gaze of Eli Ives. Next, Eaton returned to his alma mater. Williams College—situated close to the boundary point where Massachusetts, New York, and Vermont all meet—had invited Eaton to deliver a course of lectures on botany, mineralogy, and geology to any students who might be interested in these elective subjects during the spring term beginning in March 1817. Chester

Dewey, then the sole professor of mathematics and natural history at Williams, gave Eaton (a man eight years his senior) a collegial welcome. This appointment was the essential test Eaton needed to pass in order to leave his former life behind and set out on a new career with confidence and success.

Although the position at Williams was a temporary one, Eaton made the most of it. He would later boast how popular his courses at Williams were, claiming that his students became so possessed with "an uncontrollable enthusiasm for Natural History" that the other departments of learning were in danger of being "crowded out."[13] Aside from being a brilliant lecturer, Eaton possessed a commanding physique. He was about five feet, ten inches tall and weighed about two hundred pounds. Rugged rather than corpulent, Eaton's muscular body was capped by an animated face featuring fiery blue eyes and an impressively high forehead, whose outline he fastidiously maintained by shaving his fine graying black hair.[14] By all reports, he was a magnetic public speaker.

Eaton used this opportunity to carve out his own signature style of engaged science pedagogy. From the manual his students assembled out of his botany lectures, we know that Eaton would regularly take them out of the classroom and directly into the field, conducting his orientation and instruction more or less on the run like an enthusiastic nature guide. Most importantly, the conversation was not unidirectional. Eaton insisted that the students collect their own specimens of plants and then he would subject them on the spot to systematic and practical interrogations. Here, for example, Eaton presents what a model conversation might sound like when a well-prepared student has picked up a common apple blossom. Note his emphasis on the traditional Linnaean nomenclature:

Teacher: To what class does it belong?	Pupil: Icosandria.
T: Why?	P: It has 20 or more stamens fixed on the calyx.
T: What order?	P: Pentagynia.
T: Why?	P: It has 5 styles.
T: To what genus?	P: Pyrus.
T: Why?	P: It has a 5-cleft superior calyx; coral 5-petalled; pome 5-celled; each cell about 2-seeded.
T: What species is it?	P: Malus.
T: Why?	P: The flowers are insessile umbels; leaves ovate, serrate.

Eaton closes this brief vignette with a rejoinder to those who might question either its decorum or its utility: "Though the lecturer's chair is more dignified than such schoolmaster-like employment, yet the pupils will derive more benefit from a season spent in this way, and in collecting and preserving plants, than from a half-dozen courses of formal lectures."[15]

For the first time in his life, circumstances and career choice proved to be well aligned for the middle-aged, novice teacher. Banking on his interactive mode of teaching, and probably realizing it might enhance opportunities for replicating the personal bond he had enjoyed with John Torrey, Eaton began to inspire his first significant geological disciples. Though originally intent on a career as a medical doctor, Ebenezer Emmons resumed his studies with Eaton in Troy in 1826, published a manual on New York mineralogy and geology under Eaton's direction the same year, initiated his own annual lectureship in chemistry at Williams College in 1828, and then rejoined Eaton as junior professor of mineralogy and geology at the Rensselaer School in 1830. When the New York Natural History Survey was finally organized in 1836, Emmons would be selected to serve as principal geologist responsible for a major portion of the large state's territory.

During the spring recess, Eaton took creative advantage of his students' plans to visit their homes across New England, New York, and Pennsylvania. He assigned them to gather and bring back to him specimens and information about the various locations of plant and rock types. On the basis of these materials, Eaton compiled the first edition of his *Index to the Geology of the Northern States*.[16] He also revised the manuscript for his *Manual of Botany*, which he had begun in prison, substantially completed before his studies at Yale, and then tried using with his class at Williams.[17] Both volumes were ambitiously comprehensive in their territorial reach and in their degree of specificity about specimen locations and types. According to historian Joseph Ewan, "Eaton's *Manual* was *the* field reference book for every botany student in the higher schools and academies of the time, the forerunner of *Gray's Manual*."[18]

On the strength of his growing reputation, Eaton traveled eastward from Williamstown late in the summer of 1817 to begin giving a series of public lectures at Northampton, Massachusetts. Accompanied by some of his students who had just graduated, he tramped about two hundred miles in a circuitous route, taking the time to examine the Berkshire Mountains of western Massachusetts in great detail, sketching a map of territory fifteen miles wide, and saving money by staying at his students' residences when possible.[19] Eaton spent the next six months lecturing in towns across central Massachusetts, securing his social status by impressing the members of the leading families of each community with

his botanical, mineralogical, chemical, and geological demonstrations. Minis-
ters, lawyers, and doctors gathered to witness Eaton's elucidation of theoretical
knowledge through the medium of practical common sense. When Amherst's
Reverend Edward Hitchcock attended one of Eaton's presentations, the scientific
career of another prominent figure in American geology and higher education
was set in motion.[20]

Delivering a popular lecture in science in the early nineteenth century was
neither trivial nor superfluous. Historian Dirk Struik notes that many of the newer
discoveries in the natural sciences had not yet widely permeated American college
curricula or libraries, so the task of informing educated lay citizens was largely
left in the hands of traveling demonstrators like Eaton: "His method perhaps was
crude, but it answered a growing demand of the community. When Eaton ceased
lecturing in New England, others took his place." Struik quotes a letter Eaton
wrote to Torrey, sharing his advice about how to solidify one's market share as a
public lecturer: "Never offer less than 24 lectures for a course. To save the public
from imposition, make the fact known, as extensively as possible; that none but
imposters will offer less than 15 lectures for a course; and that 30 ought to be given.
These peddling swindlers, who offer to sell tickets for isolated lectures, ought
to be despised. They are always contemptible quacks of no integrity; and they
ought not to be allowed to sleep near traveler's baggage, and public inns. [These]
swindlers . . . are chiefly foreigners, and most of them are illiterate Scotch, Irish,
or English. But justice demands that we make many honorable exceptions."[21] As
he enjoyed an intimate friendship with Torrey, Eaton focused more in this letter
on giving rhetorical strategies for how to disparage the competition than on sharing any of his thoughts about effective science pedagogy.

Interestingly, if the aim of these talks was to advance the cause of original
scientific achievement in America, historians have pointed out that Eaton's commitment to lyceum-style public education was not a widely successful model. In
particular, Sally Gregory Kohlstedt notes that it was intellectually and physically
taxing to simplify scientific knowledge enough to compose intelligible public
lectures. (Eaton gave more than three thousand performances outside of institutional settings.) As professional science continued to advance, this translation
process from technical to popular explanation became increasingly difficult to
accomplish. So, she concludes, it was only reasonable for the next generation of
serious researchers to abdicate the distracting and counterproductive burden of
bringing science education to the ignorant masses.[22]

But Eaton's social agenda for science was clearly broader than Kohlstedt's narrow professional advancement premise allows. Remarkably open-minded for his

time, Eaton believed that women could be equal participants in general scientific discourse, and he used his public lectures to recruit female converts to the study of natural history. On 24 November 1817, a group of prominent Northampton citizens (including Congressman Elijah H. Mills and the former Massachusetts governor Caleb Strong) signed the following glowing endorsement: "Mr. Amos Eaton was employed in this town to deliver a course of lectures on Botany, and a course of evening lectures [on] Chemistry, Mineralogy, and Geology. As his class consisted chiefly of ladies, and as these branches of learning have not hitherto generally engaged the attention of that sex, we take the liberty to state that from this experiment we feel authorized to recommend these branches as a very useful part of female education."[23] Mary Lyon, a young student from Ashfield Academy, attended one of Eaton's chemical demonstrations in Northampton. Eaton made such a positive impression that she went on to study the natural sciences with Edward Hitchcock at Amherst College. In 1825 she spent several months with Eaton and his family, gathering valuable expertise, apparatus, and insight as she watched him prepare to open his own technical institute.[24] A dozen years later, she opened Mount Holyoke Female Seminary.[25]

Lyon was not the only prominent female educator Eaton influenced. Another remarkable woman, Almira Hart Lincoln Phelps, used his ubiquitous *Manual of Botany* as the basis for her 1829 *Familiar Lectures in Botany*. This derivative work by "Mrs. Lincoln" provided a text geared specifically toward that first generation of American women who enrolled in "female seminaries."[26] Botanical historian Howard Evans adds that "Mrs. Lincoln, like Eaton, spoke of the parts of flowers while carefully avoiding mentioning that they were sexual organs analogous to those of animals."[27] (Later, when Torrey and others of the next generation of leading American biologists were ready to discard Eaton's Linnaean "sexual system" in favor of the new "natural system" of plant classification, Eaton objected, perhaps in part because he had worked so hard to euphemize Linnaean terminology so as to make it acceptable for mixed-gender audiences.) DeWitt Clinton, ever on the prowl for intellectual society, invited Phelps and her sister Emma Hart Willard to relocate from New England to New York State in 1819.[28] They established the Troy Female Seminary in 1821, and with Eaton's active involvement and support, the natural sciences became a regular component of the curriculum at what came to be known as Emma Willard's school.

By 1817, Eaton had substantially achieved whatever original contributions he would make to the field of American botany. From 1817 onward, his creative energies shifted toward setting American mineralogy and geology on a more secure, coherent footing. Eaton spent much of his free time exploring every locale where he was invited to speak, gathering specimens and detailed knowledge of nearby geological features. Traveling across the state as far east as Worcester, Eaton bolstered his scientific expertise by acquiring a better understanding of the intervening "primitive" rocks that characterized so much of New England's terrain. The resulting revised edition of his geology text was almost six times as long as the fifty-two-page first edition, and Eaton hoped his empirically robust and rhetorically improved masterwork would better withstand the withering gaze of adversarial reviewers.[29] Taking an approach reminiscent of Thomas Jefferson's riposte to the Comte de Buffon's theory of New World degeneracy thirty-five years earlier, Eaton began to adopt a vigorously independent attitude toward expertise about American natural history. Extrapolating from the rewards of his field-based teaching approach, Eaton also began to assert a general model for the conduct of scientific research, insisting upon the importance of engaging natural phenomena through direct contact in the field.

The locus of scientific authority was primarily what was at stake. Eaton wanted to be informed about the distant pronouncements of European textbooks, but he was also fully prepared to transcend them. He understood the need for established mineral cabinets at various centers of learning, but he was reluctant to grant primacy to armchair masters of their dusty contents. In other words, Eaton began to preach a hyper-empirical mode of self-reliance in geology. He questioned every incidental repository of scientific reputation, whether it was an educational institution or a national society, and whether it had a voluminous collection of specimens gathered through agents or possessed an advantageous institutional or social position. According to Eaton, nothing could compare with the hard-won knowledge gained through extensive travel and personal eyewitness experiences. Writing to Torrey about the first edition of the *Index to the Geology of the Northern States*, Eaton anticipated the criticism that his fiercely independent attitude was likely to provoke: "I suppose you received my Index to Geology. I do not expect to have justice done me, respecting that pamphlet by any person in N. York, but Doct. Mitchill and yourself. The labor was immense; but I shall be censured for not following European geologists in a more servile manner. To have done this I must have shut my eyes to truth and common sense."[30]

Eaton's intellectual insecurity was compounded by the cultural supremacy that his contemporaries ordinarily accorded to all things European. Tellingly, although his first published scientific observations had been botanical, it was in geology that he now anchored his growing sense of unassailable confidence: "I do not like to correspond with any European excepting the geologists. Here I feel strong and *here only*."[31] Eaton's insistence on firsthand encounters with the botany, geology, and mineralogy of remote places posed one potentially insurmountable problem, however. He had been explicitly forbidden by legal statute to examine the particularly interesting geological locales of New York, his home state. If not for the twists of political fortune, Eaton might have died a tragically frustrated man.

EXILE ENDED

In the fall of 1816 Governor Daniel Tompkins won an easy reelection, but the result was rendered moot when President Madison's successor James Monroe selected Tompkins to be his vice president. In the special election that was held to fill the gubernatorial vacancy in Albany in 1817, DeWitt Clinton was favored by more than 98 percent of New York's voters. This overwhelming mandate made it possible, among many other things, for one small obstruction to the progress of science to be cleared away.[32] On 15 September 1817, Governor Clinton entered into the legal record the following note: "Whereas it is represented to us that the said Amos Eaton since that period [of his release] has devoted himself to the Instruction of youth and the Cultivation of Science and that he is a fit object of our mercy[,] Therefore know ye that we have pardoned & do by these presents pardon him unconditionally and absolutely from the felony aforesaid and of and from all sentence judgements and executions thereon."[33]

Eaton's "rehabilitation" proceeded rapidly. John Torrey immediately nominated Eaton, still busy lecturing in Massachusetts, to become a corresponding member of the Lyceum of Natural History of New York. Within just a few days, he was accepted on 22 September 1817. By the following April, Eaton was invited to give a course of lectures on natural history at Albany. DeWitt Clinton's name headed the subscription list. Eaton's authority as a speaker on natural history soon expanded beyond occasional public service events. By November 1818, Eaton had relocated his family to Troy, just upstream and across the Hudson River from Albany, and he was enjoying a steady stream of lucrative lecture engagements through the Troy Lyceum of Natural History. When this group incorporated a year later, Eaton was named its official lecturer, and through it he became closely acquainted with Congressman John D. Dickinson and the future governor William L. Marcy.[34]

Amos Eaton had returned to New York State completely transformed, a self-confident naturalist prepared to assert himself as a leader in the local scientific community. For example, when John Torrey was considering an invitation to join Major Stephen Long's proposed scientific expedition to the interior of North America to search for the headwaters of the Platte, Arkansas, and Red Rivers,

Eaton arrogantly advised his protégé: "I must talk geology to you befor[e] you see Rocky Mts. McClure [sic] is wrong in some things. I wish you to understand my Geological whims thoroughly, before you set out. I tell you—Pause a little! But I tell you—Yes, I will go through it—I tell you I am the best *practical* geologist in the U.S.!!!!!!!!!!!!!"[35]

Furthermore, by virtue of his advantageous political connections, Eaton was able to ascend quickly among the ranks of elite scientific society. The leaders of Albany's scientific community at this time were three brothers who had studied medicine, chemistry, and the various branches of natural history with Professor Samuel Mitchill and Dr. David Hosack at New York's College of Physicians and Surgeons.[36] Drs. John Beck, Lewis Beck, and Theodoric Romeyn Beck were thus important men for Eaton to impress. At the helpful urging of his friend and sponsor Governor Clinton, Eaton met first with Romeyn Beck, then principal of the Albany Academy. When Beck told Eaton about his efforts to promote the formation of a new State Agricultural Society, Eaton responded with equal enthusiasm. It proved to be yet another critical opportunity. Leading New York politicians of all stripes were behind this project. Besides Governor Clinton, the venerable surveyor general Simeon De Witt and canal commissioner Samuel Young were avid supporters of Beck's plans. Most importantly, however, Stephen Van Rensselaer was slated to be the first president of the New York Agricultural Society. Even though the Patroon had been Governor Clinton's rival, as a fellow prominent Freemason and veteran canal commissioner, Van Rensselaer was the one man whose personal wealth and enlightened politics might make it possible for Eaton to implement his innovative educational ideas and his audacious vision of an autonomous American branch of applied natural history.

Because Eaton had cultivated such a charismatic reputation as a public scientific lecturer, the Agricultural Society's founders invited him to deliver a set of lectures before the members of the New York State legislature in April 1819. The series promised to explain how the sciences of geology and chemistry might be applied to the practice of agriculture. Here was the culmination of Eaton's triumphal return to the capital of the state where he had once been maligned, wrongly imprisoned, and cast out. Eaton now displayed his newfound sense of masterful public performance, bolstered by the confidence he had derived from his strenuous scientific labors: "I do not know a person in the world, but myself, who would become a successful scientific pedlar [sic]. I have learned to act in such a polymorphous character, that I am, to men of science a curiosity, to ladies a clever schoolmaster, to old women a wizzard, to blackguards and boys a shewman and to sage legislators a *very knowing man*."[37] Every speaking engagement offered an opportunity to expand his reputation. Throughout the remainder of 1819, Eaton traveled among the major towns along the Hudson River from Troy to Catskill, giving four to eight lecture performances per week, and tirelessly collecting new specimens and impressions of the rocks and landscape he encountered.

Eaton had been willing for some time to challenge the accuracy of the pioneering work done by the few other Americans who had published geological reports of the United States. Even though he gave respectful lip service to his predecessors' accomplishments, Eaton proclaimed that the next level of understanding would require a new kind of intimate and exhaustive examination of the interior of the countryside, prosecuted at a much higher level of stratigraphic detail: "William Maclure, Esq. has already struck the grand outlines of American geographical geology. . . . Professor Mitchill has amassed a large store of materials, and annexed them to the labors of Cuvier and Jameson. But the drudgery of climbing cliffs and descending into fissures and caverns and of traversing in all directions our most rugged mountainous districts to ascertain the distinctive characters, number, and direction of our strata has devolved upon me."[38] By assigning himself this paramount responsibility, Eaton dismissed all competitors. Privately, he was boasting that he had advanced even beyond the ken of his teacher Benjamin Silliman: "He is an excellent practical chemist and a good *cabinet* mineralogist. With very little knowledge of geology, he affects much. He is impatient when his opinions are questioned, and has very lofty conceptions attached to the stupendous title of Professor of Yale College; and expects us to be ever mindful of his honorable marriage into the family of Governor Trumbull."[39]

The Amos Eaton who approached Stephen Van Rensselaer, seeking a patron for his research in 1819, was totally confident. Prepared to set American geology on a fundamentally independent and inductively sound theoretical basis, Eaton confided to Torrey that he felt that his own experience and capabilities were now unmatched: "My 2d Edition [of the *Index to Geology*] may be somewhat interesting to you. I shall be very headstrong in that; because I have no confidence in any geologist in the United States. If you would give me a written discharge from all imputation of vanity, I would say, that I am now so familiar with the rocks of New England and all that part of the state of N. York, which lies to the north of the Highlands, that I am in no danger of erring in matters of fact. I have given tone to Geology in the interior, mostly upon principles not well according with the view of our cabinet geologists; and I have hitherto seen no cause to regret it."[40]

Eaton needed just two things to gain a truly secure basis for his research. One was a title that adequately reflected his growing prowess as a lecturer on scientific subjects. He ached to hear and to see the words "Professor Eaton," used in reference to himself.[41] He also needed Stephen Van Rensselaer to bankroll his grand vision of a robust American geological science. Eaton did not at first guess that the Patroon would eventually provide both the professorship and the research support. Instead, Eaton banked on the publicity of his traveling lectures and the distribution of the second edition of his *Manual of Botany*. In May 1820, he was invited to give a six-week course of lectures as professor of Botany at Vermont's Castleton Medical Academy (then affiliated with Middlebury College). Here was a modicum of that academic status he so craved. On the eve of his departure for Castleton, Eaton received extremes of good and bad news: the second edition of his *Index of the Geology of the Northern States* was finally published, but his house, along with the whole city of Troy, was consumed by flames. Fortunately, because he had a habit of lending books and sending specimens to friends like Torrey and because he had housed others in the Troy Lyceum, a substantial portion of Eaton's private collections survived.

Torrey wrote to Eaton a month later the *Index* was already provoking some negative critical reactions. In a lengthy reply, Eaton defended his book, asserting that regardless of its defects, it was still far superior to any other: "I have now ascertained, to my full satisfaction, that I am the only person in North America, capable of judging of rock strata. Silliman does not know how to distinguish the *old red sandstone* from the more recent (breccia), nor *puddingstone* from breccia, nor greywacke from greenstone trap. At least he has committed most horrible mistakes in all these cases. On the whole he is no geologist at all, neither do I know one."[42] Eaton conceded that Maclure's book was probably the best alternative among American works but added that the difference between the two books was apparent on empirical grounds. "No person ought to write a syllable on geology, until he has seen the rock he speaks of in fifty or a hundred localities and compared all its various appearances and the various heads embraced in it." In keeping with his sense of the self-corrective function of scientific research, Eaton averred affectionate pride in the independence of his *Index*, imperfections and all: "I have come to a full determination to drive at geology in my own way. When I detect an error of my own, I will publish it, let who will laugh. And I will yield to no authority whatever, until I see the proof. I know this is talking in an arrogant manner—let it be so—I shall so proceed though I shall not talk of myself so largely excepting to *you*, because it is not fashionable."[43]

Four years later, Torrey made the mistake of reiterating his unkind remark that Eaton must be embarrassed by errors he had once propounded in print, wishing perhaps "the geology of the Northern States in purgatory." Eaton replied with a redoubled sense of shocked disbelief and self-vindication. Even though some ideas had required retraction or revision, he felt that the fruitful seeds for many others' geological work had been first established by the observations and original conjectures he had put forward:

> So far from regretting the publication of the two editions of my geology, no act of my life is a source of so much gratification to me. . . . Conybeare's theory of Deluvium and Alluvium, which is now puffed off as new, was fully developed as founded on a letter from Schoolcraft in my chapter on alluvial formations, etc. etc. two years before any Englishman ever thought of it. My analysis of the Lake Erie strontian was printed in it; so that it was in two or three hundred hands before any other persons had suspected it was not barytes. Through it I gave the first notice that [the] Catskill and Alleghany mountains were graywacke. No individual in N. York, Philadelphia, N. Haven or Boston knew what gray-wacke was, till that account appeared. On the whole, what faculty or fascicle of Spurzheim has been so severely compressed in your head of late, as to induce you to suppose I regretted the publication of that book?[44]

Lightheartedly teasing his young botanist friend with allusions to a then-famous German phrenologist, Eaton was serious about his own prospects. Having survived a series of spectacular blows (failure in his chosen profession, years of imprisonment, the deaths of his first two wives and a child he never got to meet), Eaton had grown much stronger as a person during his early forties. The wear and tear of all those painful experiences could not be effaced from his aging visage, but an absolute conviction in his own prowess as a scientist invigorated him to press harder rather than to retreat from life's obstacles. In Eaton's mind, all of his life up until 1820 had been merely the prologue to his achievement of greatness. Subsequent generations of American geologists and historians would concur, and the 1820s were remembered as the "Eatonian era" of American geology for well over a century.[45] But the circumstances required for Eaton to fulfill this remarkable destiny could only fall into place once the landlord and the ex-convict finally had the chance to be introduced.

PART TWO

Engineering for a
New World's Geology

CHAPTER FOUR

Clinton's Ditch

Before we quit the subject of the western waters, we will take a
view of their principal connexions [sic] with the Atlantic. These are
three; the Hudson's river, the Patowmac, and the Missisipi itself.
Down the last will pass all heavy commodities. But . . . it is thought
probable that European merchandize [sic] will not return through
that channel. . . . There will therefore a competition between the
Hudson and Patowmac rivers for the residue of the commerce of
all country westward of Lake Erie, on the waters of the lakes, of the
Ohio, and upper parts of the Missisipi.

— *Thomas Jefferson, 1785*

As it had begun to do in Britain, and to a lesser degree in France, the canal-building craze in the United States during the 1820s triggered major innovations in theoretical geology. Geological research would prove invaluable as an aid to the industrial and commercial innovations required to complete the Erie Canal project. By turning over a shovelful of central New York turf on 4 July 1817, De-Witt Clinton set into motion a chain of events that would allow the man he'd first met at Newgate Prison, Amos Eaton, to achieve wide recognition for compiling a systematic view of Earth's geological history. Clinton and Stephen Van Rensselaer had supreme confidence in the relevance of scientific inquiry, and throughout the next decade they would harness their respective aptitudes for public power and private wealth in parallel efforts to advance the cause of natural history. They generously sponsored a wide variety of intellectual and practical activities, including enlisting public support for the construction of the Erie Canal, and thereby made it necessary to cultivate new scientific and engineering capabilities. All of these circumstances helped to reaffirm Clinton's assumption that the ultimate social purpose of the natural sciences was to render the land's natural resources accessible, intelligible, and useful.

THE ALLURE OF INTERNAL IMPROVEMENTS

Early national leaders had long been concerned about the challenges that natural geographic barriers placed upon the health and prospects of a continent-sized republic. No other country in the world had ever attempted to function democratically over such an expanse of territory. According to leading political theorists of the French Enlightenment, who were otherwise generally enthusiastic about the American experiment in independent self-rule, history had warned that the success and stability of representative democracies was confined to localized havens of liberty, whereas larger empires were prone to succumb to despotism.[1] By this eighteenth-century conventional wisdom, if the United States wanted to avoid lapsing into the forms of social inequality and political tyranny that had provoked their struggle for independence in the first place, then its best hope lay in letting the states function autonomously as compatible sister republics within a weak confederacy. James Madison had to combat this assumption in his famous essay Federalist No. 10, which argued that a representative democracy could indeed be extended over a large territory, provided that the diversity of circumstances and interests contained within the country were balanced by a cohesion born of common national identity and fraternal affiliation.

Federalists imagined that there might be a technical solution to the problem of knitting together a scattered population. George Washington, for example, grasped the potential of an inland network of river transportation. Writing to the Marquis de Chastellux after taking a tour of the Mohawk valley in 1783, Washington confided: "I could not help taking a more contemplative and extensive view of the vast inland navigation of these United States, and could not but be struck with the immense diffusion and importance of it; and with the goodness of that Providence which has dealt his favors to us with so profuse a hand. Would to God we have the wisdom enough to improve them!"[2] Historian James Dilts posits that, as early as 1784, "Washington had foreseen the routes that were to become New York's Erie Canal, Pennsylvania's Main Line of Internal Improvements, Maryland's National Road, and Washington D. C.'s Chesapeake and Ohio Canal."[3]

Internal improvements may indeed have been in President Washington's mind, and they surely came to absorb Thomas Jefferson's attention when his turn came to serve in that high office. The political obstacles to transcending state boundaries in order to create a national transportation infrastructure were as problematic for the Republican strict constructionist as were the mountain ranges themselves. In April 1808, Jefferson's secretary of the treasury issued a comprehensive report on the young nation's prospects for constructing roads and canals. The author of the report, Albert Gallatin, was another prominent member of Philadelphia's American Philosophical Society. Gallatin saw the entire question of national "internal improvements" as being held hostage by local and regional animosities, and he sought to compile a body of public knowledge for systematic national planning.[4]

The results of Gallatin's analysis—formulated in consultation with the country's leading engineer, Benjamin Latrobe—focused primarily on securing means of communication and transportation along the Atlantic seaboard, as well as assessing the practicality of various schemes to unite seashore points to hinterlands. The largest single internal improvement task outlined in the report was a coastal inland waterway, which was to be mirrored by the completion of a post road from Maine to Georgia (a basis for what would eventually become U.S. Route 1). Gallatin's report also summarized many of the potential public road, waterway, and river improvement plans that Washington had envisioned, with special emphasis given to specific bypasses around critical natural obstacles to navigation like Niagara Falls and the Falls of the Ohio at Louisville. Latrobe also contributed reports on promising novelties such as railroad technology and steam-driven vessels. In conclusion, however, the treasury secretary deployed hard scientific analysis in an effort to discourage talk of "artificial rivers" crossing American mountain ranges. Lock navigation, Gallatin argued, would face insurmountable obstacles in the Appalachian Mountains because it "requires a greater supply of water in proportion to the height to be ascended" than any of the proximate natural bodies of water would be able to provide.[5]

Jefferson agreed with these conservative conclusions, having himself traveled along and studied the world's longest canal more than twenty years earlier. The Languedoc Canal, a 144-mile canal in southern France, had been opened in 1681 to connect the Atlantic Ocean with the Mediterranean Sea. Using tunnels, aqueducts, and 119 locks, this fantastic product of seventeenth-century engineering traversed that distance by surmounting an elevation whose highpoint was six hundred feet above sea level.[6] The "Allegheny mountains," as Gallatin had noted, were five times as high. Only where the Mohawk River opened onto the Hudson at Troy, New York, did tidal navigation approach within thirty miles of the Atlantic Ocean's watershed barrier. Since Lake Ontario was only four hundred feet higher in elevation than Troy, however, even the skeptical Gallatin was forced to concede the possibility of building a canal connecting the seaboard to the continental interior at that one point.

Few among the grand projects listed in Gallatin's plan, however, were likely to be implemented soon, given the tumult of the Napoleonic Wars, the protracted economic downturn resulting from Jefferson's Embargo Acts, and the intense local jealousies aroused whenever Congress took up the question of funding any particular regional improvement project. Even so, Federalists in New York sought congressional help to build a canal. Joshua Forman, a legislator from the central part of the state, gathered a coalition of his fellow assemblymen to pass a resolution "to direct a survey to be made of the most eligible and direct route of a canal to open a communication between the tide water of Hudson's River and Lake Erie."[7] A delegation was then dispatched to the nation's capital in the waning days of Jefferson's lame-duck administration to speak with the president about securing federal funds for the project. Jefferson dismissed the Albany petitioners with a tinge of admiration; their wonderful plan had merit, but he predicted it would not be practicable for a century.

Disappointed but undaunted, pro-canal Federalists Jonas Platt and Thomas Eddy regrouped. In February 1810, they recruited Lieutenant Governor DeWitt Clinton to join them because he was known as a man "of varied interests, a perpetual scholar whose studies range[d] from natural history to science and the arts."[8] As soon as he consented to join the commission, Clinton characteristically began to delve deeply into research on canals: consulting European engineers, reading books on European waterways, obtaining surveyors' maps of New York State, and reading a series of essays on internal navigation that had been published in 1807 in the *Ontario Messenger* under the pseudonym "Hercules."[9] Some contemporaries suspected that Clinton himself was secretly the author of this fantastic scheme to dig an overland route for navigation from Lake Erie to the Mohawk River. But the visionary originally responsible for conceiving and publicly promoting the mercantile prospects of connecting Great Lakes commerce to the Hudson was Jesse Hawley, a flour merchant from Geneva, New York, then serving a twenty-four-month sentence in the Ontario County debtor's prison at Canandaigua.[10]

THE CANAL COMMISSION

Gouverneur Morris, the former U.S. minister to France, Clinton's former colleague in the U.S. Senate, and New York's senior Federalist statesman, was chosen in 1810 to lead the Canal Commission. Five other prominent New Yorkers were chosen to join Clinton and Morris: Simeon De Witt, Thomas Eddy, William North, Peter B. Porter, and Stephen Van Rensselaer. These wealthy, prominent men represented distinct political commitments, but each also had personal, public, mercantile, or scientific interests in the proposed canal route.[11] Simeon De Witt was DeWitt Clinton's elder cousin, a man whose impressive accomplishments as the state's surveyor general after the Revolutionary War included the design and implementation of a grid system to transform former Iroquois tribal lands into parcels for veterans in the classically named military townships that now grace central New York State. Thomas Eddy, a director of the Western Inland Lock Navigation Company, was the commission's only nonpolitician and possessed some of the most directly relevant technical experience. Eddy's company had already made several attempts to create a navigable channel along the Mohawk River between Utica and Rome. General William North was a venerable Revolutionary War veteran, a staunch Federalist, and a large landowner. Peter B. Porter was a transplanted New Englander who had established a frontier law practice at Black Rock on the Niagara River. As an outspoken Republican in the U.S. House of Representatives, Porter nevertheless advocated radical maneuvers such as earmarking the sale of public lands in order to support internal improvements. Stephen Van Rensselaer was appointed in deference to his emerging interest in public works and to ensure a reasonable balance of partisan influences on the commission; at the time he was serving as a Federalist in the New York State Assembly. Finally, James Geddes, a surveyor from western New York who had made his own preliminary investigation of the terrain in 1808, was assigned to assist with logistics and documentation.

The Canal Commission's tour across New York State in the summer of 1810 allowed Clinton a rare opportunity to delve into his state's natural history with untempered zeal. His diary and the scientific notes he gathered on this tour would form the basis for most of his subsequent scientific publications (including the Hibernicus letters on the natural history and politics of New York). Among his colleagues on the Canal Commission, Clinton discovered that the Patroon was nearly his equal in terms of possessing an intellectual curiosity about nature. In that summer full of exploration and investigation, Van Rensselaer could not have foreseen that Niagara's Queenston Heights would soon be the scene of an ignominious military defeat. He was, of course, Clinton's political rival, but the two men nurtured a growing mutual respect as they traveled together across the common ground of their home state. The fever dream of an Erie Canal, and all it might bring in terms of prosperity and natural knowledge, soon infected all seven of these prominent men, and their shared experience inoculated them against the poisons of political partisanship in the future when one or another gained the upper hand in state government. To become a reality, the canal absolutely required bipartisan collaboration.

The social payoff for developing American scientific capability, in DeWitt Clinton's emerging understanding, would be the rendering of immense wealth from previously untapped land and natural resources. Physical geography spelled

out New York's natural advantages, but given Jefferson's inability and unwillingness to make the Erie Canal a national priority, it remained largely in the hands
of New Yorkers to improve the physical infrastructure. Geological processes (or
God, for the many who conceived and spoke of these matters in the providential
mode) had apparently chosen New York as the first place where nature could be
improved to meet the transportation needs of the early American republic.

On 30 December 1815, Clinton hosted a big rally in New York (where he again
held the office of mayor) to convene canal supporters and persuade opponents.
Clinton estimated that it would require $6 million to build a sixty-two-lock canal
that promised to "convey more riches on its waters than any other canal in the
world."[12] When the state legislature took up debate on the proposal in late March
1816, legislators from Long Island and towns along the Hudson were skeptical
about the economic impact of a project that would draw commercial traffic to
the central and western parts of New York. Now that Aaron Burr was disgraced,
Clinton's chief opponent among New York Republicans was the brilliant lawyer
and state senator Martin Van Buren. By stoking the fires of disaffection in parts of
the state remote from the canal route, Van Buren managed to limit legislative approval in April to a plan for construction of just the middle section, which would
connect Rome in the east with the Seneca River in the west at a cost not to exceed
$2 million. Before they sent this measure to the Senate, the assembly approved
an amendment requiring communities within twenty-five miles of the proposed
canal to pay an additional tax to support the costs of construction; by doing so,
the assembly hoped to collect the needed revenue and cajole the acquiescence of
reluctant legislators from areas remote from the canal route.[13] Playing up the un-
certainties about the practicality of Clinton's full-fledged plan, Van Buren urged
his fellow state senators to delay authorization of the canal project until the entire
route could be resurveyed during the summer of 1816. But for some delicate jock-
eying among canal supporters in both houses, failure to achieve concurrence between the two bills might have doomed the 1816 measure from ever being passed
into law.

To address the concerns Van Buren had aroused among his fellow legislators, a
newly organized Canal Commission included men chosen to represent the three
geographical regions to be served by the project: Joseph Ellicott, Myron Holley,
and Samuel Young. (Clinton and Van Rensselaer were both reappointed.) Young
was a state senator from Saratoga, representing the eastern part of the state. Holley was also a member of the legislature, but his involvement in the canal project
soon absorbed all of his talents and energy. Writing anonymously as "Tacitus,"
Clinton effusively praised Holley's work in completing the second survey of the

canal route: "This gentleman is . . . distinguished for extensive research, and acute discrimination. He has devoted his whole time and attention, mind, and body, to the canal; and some of the most luminous reports and communications have proceeded from his pen. Whatever he touches, he adorns, and whenever he speaks or writes, he instructs. His mild and conciliatory manners, his elevated character, his spotless integrity, and his indefatigable business talents, have rendered his services as an acting canal commissioner, invaluable."[14]

The third new commissioner, Joseph Ellicott, was a surveyor from western New York and younger brother to Andrew Ellicott, professor of mathematics at the U.S. military academy at West Point. It was Andrew Ellicott's documentation of the terrain at Niagara Falls that geologists later compared to Peter Kalm's earlier account in order to calculate the rate of erosion and extrapolate claims about the age of the Earth. Originally hired by the Holland Land Company in 1797, Joseph picked up where his brother left off, producing a preliminary map of the topography along the shores of Lakes Ontario and Erie by walking two hundred miles through the snowy autumn of that year. He then proceeded to lay out the 3.3 million acres of the company's purchase, arrange the specific boundaries for the Seneca Indian Reservations as provided in the 1794 Treaty of Canandaigua, and subdivide the remainder into six-mile square towns in rough accordance with the newly established federal survey practices. During all of that work, the younger Ellicott oversaw a team of 150 surveyors and instructed them to keep field notebooks containing descriptions of the land, waters, mill seats, valleys, mines, minerals, and any other information potentially useful to the company.[15] Joseph Ellicott also founded the city of Buffalo. He recognized the tremendous potential of its location beside Lake Erie and was responsible for laying out the city on its original plan. Ellicott's expert knowledge of the terrain of western New York and his deep ties to the region helped allay fears and reduce regional objections to the canal.

In the meantime, Clinton's campaigns for public support continued vigorously throughout the spring in the New York papers. In the face of a relentless and expanding tide of pro-canal rhetoric, Van Buren finally wavered, and the state legislature authorized construction of the complete project on April 15, 1817, for an estimated cost of $5 million. Since the governor's office was temporarily vacant (Tompkins had resigned, but the election for his replacement had yet to occur), a five-person council of revisions assumed veto power over the legislature's decision. That council was split evenly, with lieutenant governor John Taylor (acting as governor) and chief justice Smith Thompson strongly opposed to the canal. Chancellor James Kent remained undecided until Tompkins appeared and gave his unsolicited opinion that the state should devote all its energy to preparations for a renewal of war with the British. Kent is reputed to have proclaimed, "If we must have war, or have a canal, I am in favour of a canal," and so he sided with council members Platt and Robert Yates.[16]

In the special election to replace Governor Tompkins in the spring of 1817, no Federalist candidate could be found. Indeed, the Patroon's defeat in 1813 would prove to be the final New York Federalist campaign for the governorship. Clinton's only opposition in 1817 would have to be a Bucktail (an Anti-Clintonian Republican). Van Buren managed to persuade Peter B. Porter, the War Hawk Republican congressman and former canal commissioner from the westernmost part of the state, to run against Clinton. Voters rewarded Clinton by giving him 43,310 votes out of the 44,789 cast, but whether this could be considered an overwhelming mandate for the Erie Canal was unclear, since Porter had also expressed support for the canal. The election was clearly a mandate for Clinton.[17]

On Independence Day in 1817, the newly installed governor triumphantly thrust a shovel into the ground in Rome, New York. DeWitt Clinton understood the symbolic power of linking 4 July festivities to the inauguration of an unprecedentedly expensive and ambitious engineering effort. Besides the physical challenge of digging a navigable passage more than 350 miles long, Clinton also had to negotiate and survive the economic and political burdens of securing the public investment required to pay what turned out to be the $7 million of actual construction costs, on the promise of eventual repayment through tolls if and when the canal ever became a functioning success. Despite all the rhetoric of Providence and preordainment, there were still many choices to be made and unanticipated barriers to surmount.

AMERICA'S "FIRST SCHOOL" FOR ENGINEERS

Surveyor Andrew Ellicott insisted that Clinton not follow his first instinct, which was to send to England for engineers, and persuaded him instead to develop some home-grown talent. A chief and two assistant engineers were assigned a major section of the canal route—the western, middle, and eastern, respectively—where he would supervise the various tasks involved in canal construction. Judge Benjamin Wright, whose qualifications included surveying and mapping the length of the Mohawk valley for the Western Inland Lock Navigation Company, was chosen as the grand project's chief engineer. His reputation for settling land disputes was one reason for his selection; another was his personal connection to Federalist canal commissioner Thomas Eddy.

The supervisors recruited surveyors possessing skills with compasses, leveling instruments, and measuring chains and told them to learn hydraulic engineering. Still anxious to draw upon foreign canal expertise, Clinton sent teenaged surveyor Canvass White to England to study the principles of lock design. According to one historian of surveying, when White "returned, he brought with him new instruments, a sheaf of carefully rendered drawings, and a better knowledge of canal construction—especially locks—than any other man in America possessed."[18] White shared this critical training in the principles of lock, bridge, and dam construction with his countrymen upon his return, but formidable and costly obstacles to the grand project remained. In order to build the canal, which had to be a watertight container for vessels to pass across swampy terrain in western New York, a waterproof cement was required. At the inception of the project, tons of the only known source for this cement's crucial ingredient, a limestone quarried in Wales, were crushed and transported across the Atlantic Ocean at incredible expense. Three years into the excavation of the Erie Canal, however, a native lime rock was unearthed in central New York which produced superior hydraulic cement. Although White obtained a patent for this homegrown substitute in 1820, Clinton shared with chief engineer Benjamin Wright the honor of announcing the discovery of New York's bountiful local supply of "waterproof" limestone to the scientific world in 1821.[19]

At one stroke, a geochemical experiment had ensured that a critically needed ingredient would be cheap and plentiful. In Clinton's view, this was precisely how science was supposed to work, and more than convenience was at stake. Writing to William Darby in 1819, Isaac Briggs noted that great projects tend to surpass their initial expense estimates. For example, his eastern section of the Erie Canal would require chiseling a channel through difficult rocks and the completion of a complicated series of closely spaced locks. This part of the project was being saved until last since it was estimated that the cost per mile for construction here might be discouraging to the state legislators who had approved funding for the project. Briggs blustered hopefully to Darby, "It is my decided opinion that this portion of the canal can be made for an expense averaging 16 per cent . . . less than the [$2,700 per mile] estimate of the commissioners."[20] He was mistaken. Construction on every section ran into nearly insuperable difficulties at one point or another. In the middle, for example, near the Montezuma swamps, mosquitoes carrying deadly infectious diseases and the consistency of the swamps themselves conspired to bring progress to a standstill in the summer of 1821. Workers became ill and the channel had to be repeatedly excavated because the sides of the canal would ooze and refill each day's progress before the next day could begin. Had it not been for the windfall of the native waterproof lime, a discovery unforeseen when construction began, the state legislature might well have refused to authorize the continuation of the work in the face of these mounting unanticipated costs. Attention to local natural history permitted geological fieldwork to go hand in hand with the canal's construction, thereby rescuing the feasibility of Clinton's gigantic project. Miraculously, and only for this reason, the Erie Canal lived up to Briggs's rose-colored expense prognostications after all.

The science-minded governor drew a valuable lesson from the mounting evidence that engineering expertise could be home grown, and he acted on the inspiration that natural history expertise might be particularly relevant to the success of canal development. When Geddes's work on the western section required the appointment of another engineer, Clinton used his position as canal commissioner to hire the services of a natural history expert. David Thomas, a Quaker naturalist and specialist in botany educated in Philadelphia, lived in the hamlet of Aurora overlooking the eastern shore of Lake Cayuga. Thomas had established a botanical garden and was rapidly becoming a major figure in the network of scientific correspondents about the botany of North America. He had also recently published a journal of his travels from Aurora to the Wabash River in the Indiana Territory. Clinton purchased twelve copies of the volume, which described "the natural history, topography, geology, antiquities, manufactures, agriculture and commerce of the western country," and is said to have remarked to his fellow commissioners: "The man who wrote that book will make an excellent canal engineer."[21]

Thomas agreed to work on the construction and maintenance of the canal between Rochester and Lake Erie, though he had no relevant prior experience. Clinton, blithely confident in the skills of a natural scientist, supposed that the botanist could learn the rudiments of engineering just as quickly as frontier judges and surveyors had. He gave explicit and detailed instructions to his newest engineer to continue work as an active scientific investigator, urging Thomas to investigate the curious parallel scratches found on the limestone rocks east of the Genesee River. Clinton also expected Thomas to identify and ship to Albany any valuable mineral and fossil specimens that came to light in the canal work.

But before any lucrative or esteem-building research projects could be launched, Thomas would need to learn how to build canals. In his letter of employment dated 7 March 1820, Clinton told Thomas to read up on hydraulic theory and engineering practice: "You will receive instructions from . . . the Acting Commissioners. As to Levelling you can learn it in a day. You will have I presume a Surveyor to assist you and perhaps Mr. Holley may be able to accommodate you with a levelling instrument."[22] By August of that year, Thomas was receiving both encouragement and warning from Benjamin Wright, who had been instructing him in how to use Holley's leveling instrument:

I am glad of your progress westward and shall consider if you reach Tonawanda Creek this season and also make a little *Eye* Examination of the country between that and Buffalo it will be doing well. . . . If you see the Argus printed at Albany you will find some harsh attacks upon the Governor & Mr. Holley as I am told—I have not seen them all—I consider it is preparing the way for removal of those two gentlemen from the Board of Canal Coms. next winter. I should regret much such an event as I fear it would be the forerunner of destruction to the Canal. If Political considerations are to govern the Canal then good bye to Canal—the Funeral dirge is near at hand.[23]

By February 1821, Holley was reporting recent political victories for the canal appropriations in the state assembly but also warning that if he and Clinton were displaced he could not promise that Thomas's services would be retained.[24] The summer of 1821 proved to be very frustrating for canal construction with the mosquitoes and oozing muck blocking progress at the Montezuma swamp and rock blasting at the Hudson terminus experiencing many delays. Nevertheless, September found Governor Clinton in excellent spirits, writing with appreciation to his western district engineer and thanking him profusely for all the seeds and botanical items he had shipped to Albany thus far: "In giving you this trouble I must mention that it is to [satisfy] Grateful Savans abroad. The attention of Europe is wonderfully attracted to this Country and its productions. I will refund any expense that may occur. If you have any spare minerals for my cabinet, I will accept of them with thanks."[25] Clinton's appetite for more specimens to add to his private cabinet would continue to pepper his correspondence with Thomas.[26]

Although Holley relinquished his spot on the commission the following spring, Thomas managed to keep his job, and while political concerns continued to threaten the future of canal construction, the engineer maintained a steady traffic of scientific correspondence and a supply of interesting specimens to his patron. In July 1822, Clinton queried Thomas about ornithology; in September, the governor wrote at least twice, once to request a batch of specimens on behalf of the Horticultural Society of London and separately to ask for more botanical specimens for himself.[27] Clinton was successfully using his political powers of patronage to ensure the enlargement of his international reputation as a public intellectual.

MYSTERIOUS TOADS, CLAMS, AND ICE CAKES

The canal was a boon to the practice of natural history, for the phenomena being exposed to human inquiry were provocative, if not absolutely mystifying. Once

Amos Eaton entered into a patronage relationship with Stephen Van Rensselaer, public discourse about the curiosities unearthed in the digging of the canal stimulated a program of scrupulous field research along the Erie Canal's route (see chapter 5) and the development of a practical educational philosophy to accompany Clinton's democratic vision for the dissemination of scientific knowledge (chapter 6). Writing in 1820 to Henry Schoolcraft, a fellow New York mineralogist who had successfully launched his career as a naturalist and Indian agent in the upper Mississippi valley, Amos Eaton promised, "I intend to kindle a blaze of geological zeal before you return. I have adapted the style of my index to the capacity of ladies, plow-joggers, and mechanics."[28] New York's grand window into the Earth's history was at that moment beginning to be exposed, and as laymen become interested observers of the curiosities being uncovered, their respect for scientifically trained arbiters of knowledge also grew.

One never knew what might turn up in the shovels of the canal's excavators. Workers were digging up all kinds of bizarre things from New York's soil. A torpid toad from Lockport, for example, found its way into both the popular newspapers and the scientific journals. While passing through Buffalo on a return trip west from New York City in 1822, Schoolcraft learned about this toad, "which on exposure to the air instantly came to life, but died in a few moments afterwards."[29] Workmen who had been blasting the rock to create a place to build locks insisted that the cavity in which the toad was found was only large enough to contain the body. Six inches of solid stone intervened between the enclosure and the surface, and they swore that no avenue for movement or the exchange of air could have been available to the creature. The newspaper-reading public marveled at the toad's miraculous ability to come back to life. Schoolcraft, as a trained naturalist, was not so easily impressed: "Of the causes which enable animals of this class, which have been suddenly enveloped in strata of earth, or otherwise shut out from the air without injury to the animal organs, to resume, for a limited period, the functions of life, on being restored to the atmosphere, no explanation need here be given, as the occurrence is a very common one, and is perhaps always, more or less, the result of galvanic excitement." For Schoolcraft, it was more remarkable that the antediluvian toad's tissues had been so perfectly unharmed within the solid limestone. He saw evidence here to confirm the Neptunian views of secondary rock formation as opposed to those of the Vulcanists: "If secondary rocks, as Hutton and Playfair have taught, have been consolidated by fire, would not the animal here incarcerated, have been consumed, or at least, such an effect have been produced, upon the animal organization, as to prevent resuscitation?"[30]

Though he was a keen seeker of evidence of antediluvian remains in New York, Amos Eaton professed serious reservations about such reports, and this one in particular troubled him for several years: "The discovery of toads in secondary rocks is often announced in public journals. A very particular account was published in the newspapers, of one found at Lockport, while they were cutting the canal bed in the geodiferous lime rock. I collected all the facts in my power, and examined the rock from which it was taken. The evidence would have been sufficient to establish any ordinary fact. But there seemed to be so many ways for illiterate laborers to deceive themselves, that I took no further notice of the report."[31] Eaton may have mistrusted the testimony of illiterate canal diggers, but he did not reject the access that they provided to fresh scientific discoveries. In 1829, he published a comprehensive report on useful organic and inorganic substances "of recent geological origin" to be found beneath the surface of the soil. Here Eaton contrasted the dubious story of the torpid toad with a similarly surprising discovery made by Erie Canal diggers, who "found several hundred of live molluscous animals. 'I have before me,' says Professor Eaton, 'several of the shells from which the workmen took the animals, fried and ate them. I have received satisfactory assurances that the animals were taken alive from the depth of forty-two feet.'"[32] The conclusion Eaton drew from this episode is especially interesting, since it simultaneously demonstrates his awareness of contemporary European theories of species change and reveals his conceptual reluctance to embrace a geological time deep enough to afford evolutionary biological change. Regarding the "fresh" mollusks harvested from that diluvial rock, Eaton quipped: "At any rate, they prove the absurdity of [Erasmus] Darwin's hypothesis—that all animals are perfected at every successive generation, and that man 'probably began his career as a fish.' For those fresh water clams of three thousand years old, precisely resemble the same species which now inhabit the fresh waters of that district."[33]

Sophisticated interpretation of the fossil record would always be one of Eaton's weaknesses as a geologist, but in the realm of structural geology he was usually quick to comprehend the implications of his observations. Early in 1822, he published a remarkable report that suggests how receptive his mind would have been to the glacial hypothesis had it been available to him. As the Hudson began to thaw at the end of that winter, Eaton paid close attention to the motion of ice blocks floating downstream. He observed their interaction with the barrier created when Briggs's team of canal engineers began to build a sloop lock at the canal's prospective eastern outlet into the Hudson River.

Writing to Silliman, Eaton guessed conservatively at the power of ice to move and to elevate rocks and soil. Specifically, he documented a substantial deposit of rocks and soil whose summit rose far above the surface of the water. He also remarked that this was gravel that had been borne a great distance by hundreds of cakes of ice. Eaton's article described in detail how the ice blocks arrived at the half-finished canal sloop lock and began to pile up, pressing against each other until some of them shot right over a little rocky promontory and into the canal. This process continued until "at length an enormous ice cake appeared, bearing on its back a great quantity of gravel." The constant pressure of additional ice cakes carried by the current forced the largest cake across the canal and up the opposite bank "so that its eastern edge extended thirty-four feet higher than the surface of the water." Hundreds more of the ice cakes were eventually propelled to the same height, forming a mountain of ice which, once it melted, would leave tons of gravel from the northern counties piled high along the canal terminus. Eaton's point in documenting this episode is particularly provocative: "Since it seems to be a rule among geologists to trace the derivation of alluvial deposits to localities more elevated than those where they are found, it may be well to remind them of contingencies of the above nature."[34]

In other words, Eaton's geological observations of the canal construction project in 1822 substantiated one of the very first hints in American geology of how moving ice could possibly be responsible for transporting the gravel and stones that every geologist tended at the time to call "diluvial debris." Despite this isolated insight, neither Eaton nor his contemporaries were swayed in their general conviction about the predominance of water as a geological agent. The assignment DeWitt Clinton gave to David Thomas about those parallel scratches in western New York might have highlighted the rock-sculpting powers of ice, but the skilled naturalist was wholly unprepared to think in such terms.[35] While employed as chief engineer of the Cayuga and Seneca Canal in 1829, Thomas would finally report his findings on "Diluvial Furrows and Scratches" that he had observed at various locations and altitudes along the canal route between Lockport and the east side of the Genesee River. Thomas compared the scratches in western New York to those he had previously observed on hard rock in situ "on the Montrose and Milford turnpike, south of the Great Bend of the Susquehanna." He was as mystified as others about the regularity of the orientation of the marks but concluded only that they were evidence of a general inundation, most likely caused by Noah's flood: "I see no difficulty in referring this attrition of the surface of rocky strata, to the Deluge, — a period when all the loose matter of the globe appears to have been in violent commotion; but on the cause of the lines so regular and so deeply engraved, I have nothing to offer."[36]

These speculative writings by Eaton and Thomas, documenting classic examples of the consequences of glaciation without understanding them as such, came more than a decade before British geologist Charles Lyell took an interest in the effects produced by drifting icebergs and fifteen years before Swiss geologist and paleontologist Louis Agassiz first popularized his fantastic (but ultimately confirmed) visions of an ice age. It would require decades of debate for the world's leading geologists to accept the idea that ice could have had such a formidable land-shaping power.[37] When a doctor from western Massachusetts reported his hypothesis in 1842 that the train of boulders strewn across Berkshire County might have been produced by a gigantic glacier, Lyell was able to examine the evidence firsthand.[38] Though he was by then the world's most influential geologist, Lyell simply could not comprehend the fantastic image of a sheet of ice a mile thick spreading across a continental land mass. Greenland's interior would not be explored until more than fifteen years later, and so in deference to the scientific principle of relying only on actually observable natural processes, Lyell resolved instead to defend the theory that icebergs must have transported those massive erratic rocks. On a grander scale than Eaton's Hudson River ice blocks but acting much in the same manner, Lyell supposed that rocks embedded underneath the floating ice mountains were responsible for etching parallel scratches across the sedimentary rock surfaces, which must have formed the bottom of a shallow sea floor.

For geologists, the Erie Canal provided access to provocative clues to the history of the Earth; meanwhile, geology provided some essential information ensuring the practical possibility of completing the Erie Canal. This reciprocity can be understood as a reflection of the American version of the Baconian scientific worldview: the interrogation of nature for the sake of increasing knowledge was perfectly compatible with the exploitation of nature for the sake of increasing prosperity. DeWitt Clinton was just one particularly well-informed and well-positioned man who brought this worldview into American politics and culture, but his fellow canal commissioners, engineers, and scientific protégés were also aligned by their shared historical circumstances to help fulfill the promise of technology. Nature's gifts existed to be discovered and then utilized to meet society's perceived needs. For men like Henry Schoolcraft and Amos Eaton, science's contribution to the harvest of natural wealth was not simply a crassly materialistic endeavor. These men hoped to gain, on behalf of themselves and their country, a share of the credit to be won for resolving the mysteries of the Earth and its history. Eaton's career in Troy was about to blossom into its full potential, primarily because of Stephen Van Rensselaer's patronage and interest in the Erie Canal.

CHAPTER FIVE

Eaton's Agricultural and Geological Surveys

This being the greatest undertaking of the kind, I believe, hitherto
known in America, I feel a deep interest in it. You perceive that
my reputation is much at stake upon it. It will probably be the
most conspicuous situation that I shall ever be placed in; and the
responsibility the greatest. A strip of more than three hundred miles
in length, which is forever to be travelled over by the learned of all
countries in canal, stage-boats, etc. I am to examine minutely and
decide upon the rocks, minerals, soils, and plants.

—*Amos Eaton, 1822*

Governor DeWitt Clinton's pardon permitted Amos Eaton to return to New
York State in 1819. Eaton then devoted the remainder of his life to becoming the
preeminent lecturer and recorder of the natural history of the northern United
States. After a brief sojourn in New Haven to repair the deficiencies of his self-
education, Eaton carried glowing recommendations from New York's celebrated
Dr. Samuel Mitchill, Yale's botanist Dr. Eli Ives, and the venerated editor of the
American Journal of Science and the Arts, Professor Benjamin Silliman. Eaton was
well prepared and well situated to take advantage of a rare scientific opportunity
to investigate the geological record then being exposed by the digging of the Erie
Canal. His prominence in the Albany scientific community was vastly enhanced
when he became acquainted with the founders of the New York State Agricultural Society, all past or current members of the canal commission.

AN INVENTORY OF THE PATROON'S PROPERTY

In 1820, after lecturing in New England towns in the spring and teaching botany
at the Castleton Medical Academy in the summer, Eaton returned to Troy and
undertook his first lucrative scientific research assignments. The president of the
Agricultural Society, Stephen Van Rensselaer, had commissioned geological sur-

veys to be conducted for Albany County and Rensselaer County, both of which happened to belong to his personal estate. Eaton and Romeyn Beck were selected to conduct the first survey. Beck enlisted one of his Albany Academy students to assist in the survey work, a lad named Joseph Henry. (Henry, who would become a world-renowned experimental physicist and later serve as the first secretary of the Smithsonian Institution, initially came to the prep school to pursue his interest in poetry and drama.)[1] Eaton reveled in the opportunity to exercise his talents for geological field work and systematic organization of facts, and six weeks of tramping around Albany County in August and September 1820 yielded a highly acclaimed publication before the end of the year.

Writing from New York City in early December, Dr. Mitchill praised Beck, his former student: "I have perused with close attention the geological survey of Albany County made for the Agricultural society by yourself and Mr. Eaton." Mitchill's effusive congratulations spoke directly to the criteria of utility that Eaton had espoused in his lectures and scientific articles: "The facts . . . are so arranged and digested as to lead to useful and practical conclusions. The description of the rocks, and of the formation of soil by their disintegration and crumbling, appears to have been executed with fidelity as well as labour and skill. . . . Your publication is a great model for imitation; and I hope it will prompt other persons to engage in similar pursuits." Waxing on, Mitchill speculated: "What can be more interesting than such deductions as you have made from the analysis of arable soil, on the modes of culture and the crops to be raised? Your notices of the organic Remains [fossils], and your sketch of the Helderberg [an escarpment that defines the southeastern boundary of the Mohawk valley], are particularly agreable [sic] to me."[2]

More important than Mitchill's pleasure was Van Rensselaer's satisfaction at the result. By skillfully and sensibly rendering technical information, Eaton had avoided that brand of fanciful landscape description so popular among his contemporaries. He was a serious, practical-minded thinker about the structure, placement, and meaning of rocks, fossils, natural vegetation, and cultivated crops, and he employed just the blend of ordinary fact and sophisticated analysis best suited to appeal to Van Rensselaer's vision of the ideal landholding gentleman-citizen. Eaton's enthusiasm and apparently boundless energy suggest that he felt rejuvenated by his new career, but bouts of severely poor health during the winter of 1820–21 brought home the reality that he was not a youth like his assistant Joseph Henry but a man in his mid-forties. Doctors diagnosed his troubles as incipient typhus, followed by an attack of cynanche (severe sore throat).[3] In spite of these ailments, Eaton continued to travel, lecture, and actively assist in Troy's resurrection from the devastating fire that had begun on midsummer's eve in June 1820 and smoldered for several weeks.

By the summer of 1821, Eaton was restored to health and ready to resume geological work. The survey of Rensselaer County remained to be done, and this time he would enjoy the company of Romeyn's brother Lewis Beck. Historian Walter Hendrickson credits Eaton's Rensselaer County survey of 1821 as the direct forerunner of all later state geological surveys: "This survey had a definite economic purpose, and in addition to making a geological section and giving a lithological description, Eaton noted the kinds of soil, the most suitable crops and the best methods of cultivating them—all matters of concern to a landowner."[4] With the assistance of the Beck brothers and Joseph Henry, and with the Patroon's generous financial support, Eaton was creating a new institutional form for the practice of geology. Because Eaton had completed his first two geological survey assignments with alacrity and success, Van Rensselaer agreed in 1822 to bankroll his next proposal, which was to survey the geology and agricultural character of towns abutting the Erie Canal route—a vast region that would soon be opened up to commerce across the central and western parts of the state. Over the next six years, Eaton would parlay Van Rensselaer's initial commitment into material support for at least six research tours to examine the canal route, as well as auxiliary excursions eastward into New England and southward into the Catskill Mountains and beyond into parts of New Jersey and Pennsylvania. Eaton clearly understood how significant this charge would be for his career. New York's canal-digging endeavor provided continuous exposures of bedrock layers and a constant supply of interesting specimens. At that moment, no other place in the world offered such a wealth of opportunity for serious collecting and observation of the products of Earth's natural history.[5]

Eaton wrote to Silliman immediately upon receiving Van Rensselaer's approval to go ahead with the plan. His tone was boastful and yet cautious: "The Patroon (as we call the Hon. Stephen Van Rensselaer) has concluded to take a Geological and Agricultural Survey across our state, following the canal rout [sic]. You know he always calls on me in such cases; and I have of course, engaged to undertake it."[6] Eaton's sense of being in the right place at the right time carried the added responsibility of "getting things right" about a tract destined to be more heavily traveled and intimately inspected than virtually any other place in America. The scenery along the way west from the Mohawk River to Niagara Falls was already deeply etched in the minds of armchair voyagers who continued to read the slew of travel accounts steadily being turned out. Thousands of immigrants and visitors would surely follow the opening of the canal to commercial navigation, and each of these people could negotiate an independent understanding of the landscape through direct observation, as filtered by religious, scientific, and aesthetic considerations. By performing a comprehensive and accurate survey of the entire region, Eaton had a chance to educate more than just the casual readership of guidebooks like Mitchill's. He was poised to influence all subsequent scientific discourse about New York's natural history. Fulfillment of this promise would require that he take into account all the mysterious features that had attracted popular notice and yet move far beyond the kinds of superficial descriptions and speculations that earlier writers such as Mitchill Clinton, William Darby, and Timothy Dwight had already provided to interested readers by the early 1820s.

ROCK FORMATIONS REVEALED BY THE ERIE CANAL

Eaton set out on his first tour of the canal route late in the fall of 1822. His initial reconnaissance that November and December was an effort merely to identify the basis rock types and characters. Eaton was preoccupied with cataloguing all the bedrock layers he encountered, and he was intellectually thrilled by the challenge of developing a precise, systematic, innovative nomenclature to apply to New York's strata. (Chapter 6 examines the results of these efforts in greater detail.) It was immediately apparent to him that virtually all of the canal route west of Schenectady contained a wide variety of the kinds of rocks that William Maclure and most other Wernerians would have been content merely to classify as "secondary" or "transition rocks." Eaton noted with interest, especially in light of his eye-opening excursion into the hills near Albany with Charles Lesueur, that these sedimentary rocks to the west were abundantly populated with the fossil remains of remarkably diverse marine creatures. Equally mysteriously, the terrain was also liberally scattered with erratic boulders and gravel composed of "primary" rocks like gneiss and granite, obviously alien imports somehow brought to rest upon the local basis rocks.

Other than tracing and documenting the orderly superposition of dozens of different rock layers, which Eaton accomplished by referring primarily to their distinctive mineralogical characteristics, his most noteworthy discovery on the initial reconnaissance trip across New York State came on 19 November 1822. In Chittenango, Eaton saw a fossil tree that was "very unlike Schoolcraft's tree."[7] Henry Schoolcraft had recently made quite a splash with his 1821 report of a "remarkable" fossil tree he had encountered fifty miles southwest of Lake Michigan during an expedition led by Michigan's territorial governor Lewis Cass to explore the upper Mississippi valley. Schoolcraft's discovery was deemed so noteworthy that the three living former presidents of the United States (John Adams, Thomas

Jefferson, and James Madison) all agreed to append their own scientific opinions regarding the meaning of the fossil tree. The three retired politicians' learned responses to Schoolcraft's report appeared together as an article in *Silliman's Journal* the following year.[8] By the beginning of December 1822, Eaton returned to Troy and began busily writing up his notes from the quick inspection he had just made of the canal route. He also drew up a more comprehensive plan of investigation that indicated those localities where he would need to spend more time carefully gathering observations in order to get a reasonably complete picture of the geology of the entire district.

In his second canal tour, which occupied the spring months of May and June 1823, Eaton focused on the large boulders resting among what he called the superficial layers of "broad alluvion." He also continued to gather empirical details about the bedrock strata, but he was not yet ready to pronounce upon the geological and agricultural potential of the canal district. He insisted on retracing his steps two more times that year before settling down to prepare his first volume of the survey report he had promised Van Rensselaer. Eaton initiated his third canal tour shortly after the close of a short summer course of teaching natural history to the students at Emma Willard's Troy Female Academy. Recollecting the series of rock layers and their locations from the two tours he had made in the previous nine months, but having misplaced his actual journal of field notes from the May–June tour, Eaton started out again in July 1823. He hoped to use this trip to find answers to practical, economic concerns so that his geological report might appeal to a broader audience than people already interested in geology. In particular, he sought to establish ways of identifying which surface rock types might promise underlying supplies of salt, as well as other valuable minerals such as gypsum and coal. Salt, of course, was an important food preservative before modern refrigeration, and gypsum was one of the first mineral fertilizers (calcium sulphate) to be widely used to repair soils that had been compacted through cultivation or rendered increasingly salty through irrigation. With the advent of steam-powered manufacturing and transportation, early nineteenth-century New Yorkers were highly conscious of the state's need to import anthracite coal from nearby sources in Pennsylvania. Nothing would have delighted Eaton's patrons among the Albany elite more than a discovery of native coal.

In addition to these practical considerations, Eaton began to devote passages in his newest journal volume to ideas leading toward a comprehensive theoretical framework for understanding the structural geology of New York State. For example, the very existence of the Adirondack Mountains puzzled him greatly. This range, which Eaton identified as the McComb's [sic] Mountain formation, interrupted an otherwise continuous family of strata deriving from the Green Mountains of Vermont and extending both north into Canada and south to the Catskills.[9] Eaton reflected upon what he considered to be the notable contrasts between the two sets of formations. For example, he observed that fossils rarely appear in the Vermont hills in the same strata where they are plentiful in New York. But what really stands out is his attempt—in a geological journal mostly devoted to recording raw observational data—to develop a working model of genetic relationships among the major mountain chains of the northeastern United States: "There is some confusion at the meeting of these series. It is sometimes difficult to determine which of the two passes undermost. The Mohawk seems to be the dividing line for many miles." Eaton's mind raced ahead with an intriguing line of logical deduction. Perhaps superficial geography could provide reliable indicators of underlying geological realities: "That a river or lake should be a dividing line between strata or formations is a reasonable supposition. Depressions and chasms will of course become the beds of rivers or lakes. And depressions will more frequently occur at the meeting of different formations." He was certain, however, that his fellow students of New York geology, such as Edwin James, had mistakenly assumed continuity between the Appalachian range in New York (the Catskills) and that of Quebec (the Laurentians). Eaton's McComb series (the Adirondacks) clearly interrupted the sequence of older strata. Here he noted explicitly how a successful human endeavor to modify nature could expose useful clues to geological history: "Though [the Erie Canal] is an artificial work, its projectors assumed a natural bed for it. And it is just upon the margin of the McComb's Mountain series from a few miles west of Schenectady, at least to Genesee River. The meeting of those vast series of strata, originating from such different primitive roots, as might be expected, result in very interesting geological productions, and require most minute examinations."[10]

For Eaton, the canal was not simply a convenient and extensive slice of representative terrain. It was a critical site for geological investigation. The manner in which he sought to decipher Earth's history from the evidence he was able to gather in his early canal tours upends two assumptions often reinforced by the standard historiography of early nineteenth-century American geology: first, that the goal of "catastrophist" geology in the 1820s was to reveal God's design in Nature; second, that the traffic of natural knowledge flowed one way—from the scientist to the engineer. Eaton is usually lumped together with other Americans who adapted Cuvierian catastrophism to reconcile evidence of deep time with some viable interpretation of Christian scriptures. A close reading of his journal demonstrates, however, that Eaton looked first to everyday sources of change as the key basis for any scientific explanation of a geological formation. Indeed, his respect for the power of ordinary erosive forces puts him squarely in line with the fundamental "uniformitarian" attitude that Charles Lyell would enshrine shortly thereafter as a dogmatic principle of modern geological science: the idea that actual observable everyday processes must be sufficient to explain the formation of all terrestrial phenomena.

Regarding the second typical assumption, Eaton was unusually sensitive in recognizing and articulating a close relationship between geological preconditions and effective human modifications of nature. As one illustration of his precocious sophistication, Eaton's views on the Erie Canal's location might be compared to a late twentieth-century science writer's reflections on the same issue: "The topography and the drainage established on [the canal] are largely controlled by the type and 'structure,' or geometry of deformation, of the underlying bedrock. The Erie Canal utilized, as fully as possible, the existing east-west waterways . . . and followed their lowland belt to minimize topographic obstacles and cost, while, at the same time, bridging the vertical distance of 545 feet between Lake Erie and the Hudson River at Albany. If a more northerly or southerly route had been chosen, the canal might never have been finished."[11]

EARLY RETURNS ON THE PATROON'S INVESTMENT

During the third tour, Eaton ventured further from the actual canal route than he had on his previous journeys, in order to gather lore and data about the location of gypsum, salt, and coal in the nearby countryside. At the brim of the Great Montezuma Swamp west of Syracuse, he remarked in his journal: "Is it not probable, or in fact certain, that the strata at the brim of the great swamp basin are thicker than at the bottom? Does this indicate, that there may be other strata between the transition graywacke (such as coal, &c.) in the thin parts?"[12] He would later expand these musings into a tantalizing (but, unfortunately for Eaton, quite incorrect) claim on behalf of New York's mineral potential: "There may be millions of millions of chaldrons of coal between this rock and the graywacke rock beneath it, without our being able to hit upon a bed of it, at one, two, or even a dozen borings."[13]

Eaton's futile quest for New York coal drew him to investigate every rumor of "burning springs," which he believed to be reliable surface indicators of underlying fossil fuels. These detours sometimes lured him quite far from the canal route, but along the way Eaton gained valuable familiarity with the character of the surrounding terrain, especially the major attractions for geological speculation in this region. For example, Eaton traveled to Clarkson to observe the elevation of the Ridge Road, which parallels the southern shore of Lake Ontario. The phenomenon of the ridge was already quite well known. William Darby's account had made much of the relative altitudes of the ridge and more distant, evidently flooded lands. Eaton took his predecessors to task: "Some gentlemen of first talents, have hastily constructed very ingenious theories to account for the existence of the Ridge Road range." Clear evidence that the lake's shore had once been higher had been interpreted in an extremely providential manner. Eaton could not take seriously the implication that a "congress of aqueous particles, the predecessors of the present which constitute the lake, assembled and said 'God, let us build us' a great basin for our future residue, with a brim from which we will hereafter recede and have a dry elevation, whereon man will construct a road." Instead, he concluded that the great basin of the lake was first made by some ordinary geological process (though he still did not have an inkling of the scale of continental glaciation that was ultimately to be credited), and then the water settled down into that basin "on the common principle of water seeking its lowest level."[14]

Eaton hoped to use science to demystify the features of the western New York landscape, and so he took a distinctly skeptical approach to ideas that associated the ridge with fantastic "bursting lakes" schemes. His journal mixes observations with critical reactions to the prevailing pattern of explanation. He was predisposed to believe the lakebed was formed by some agency other than the resident body of water. By reserving his judgment on the matter, he also exercised his doubt that Nature always reflects a divine design geared for human benefit. In his published version of the canal report, he offered no theory to account for the ridge, but he did examine the underlying rock offsets and compared these to the kind of rounded, elevated steps that were also to be found in the Catskills.[15] Being in western New York, Eaton was virtually obligated to comment upon the recession of Niagara Falls. No one who practiced geology in the state could resist observing and pronouncing on this infamous locale. Even James Geddes, the engineer placed in charge of building the western section of the canal, shared his opinion about the history of Niagara Falls in a letter to Romeyn Beck which found its way into *Silliman's Journal*. Geddes argued, contrary to the opinion promoted in most published scientific texts by that time, that the falls could not have regressed all the way from Lewiston but instead had only receded backwards from the whirlpool.[16] By resetting the erosion chronometer's starting place so much closer to the current position of the falls, Geddes's view substantially enhanced the viability of a literal biblical calculation of the age of the Earth.

The third canal tour ended in late August 1823, when Eaton rushed back east to accompany Van Rensselaer to the official opening of the Champlain Canal (which opened up navigation between the Hudson River and Lake Champlain). The great exertion of the summer's travels had again left Eaton feeling quite ill, suffering inflamed lungs and fevers well into the fall. Despite his poor health, he worked on a geological map of the canal route in September 1823. Van Rensselaer had expressed his pleasure at the survey's success back in August, but he worried that Eaton's proposed changes in nomenclature would require the sanction of a scientific geologist who was "known to the public."[17] After attempting to consult directly with Silliman, Van Rensselaer wrote back in early October to warn Eaton not to expect any help from that quarter. Silliman was severely ill himself, and he had informed the Patroon that he might have to reduce his teaching duties at Yale during the upcoming term.[18] Left to his own devices and still physically unwell, Eaton mounted yet one more quick jaunt across the state as soon as his fall term of teaching botany at Vermont's Castleton Medical College was concluded. During this fourth tour, which took just a few weeks in November 1823, Eaton tried to secure the handful of uncertain details he needed to complete the map for the survey report. He stayed close to the canal, traveling by packet boat as far west as Clyde, just beyond the Montezuma swamp, at which point ice forced his return.

Eaton submitted the completed manuscript to his publisher barely eight weeks later, on 22 January 1824.[19] His patron was in Washington, having been elected in 1822 to fill the Van Rensselaers' customary seat in the U.S. House of Representatives. The opening had been created when Stephen's cousin congressman Solomon Van Rensselaer was appointed postmaster for Albany.[20] Shuttling between Albany and Washington for the rest of the decade meant that Stephen Van Rensselaer was rarely nearby when he and Eaton needed to consult. Such was the case now, as both impatiently awaited the publication of the survey. In February, the Patroon wrote from Washington, "The public expectation is raised & daily enquiries are made when I expect your report."[21] Upon its arrival in early March, Van Rensselaer expressed pride and gratitude for the return on his investment. The first part of the survey of the canal route was an impressive achievement. Within the space of just eighteen months, Eaton had traversed part or all of the Erie Canal route four times. He had established the basic outlines of the strata and, by incorporating data from Edward Hitchcock's contemporary survey of the state of Massachusetts, compiled a geological profile of the rocks stretching all the way from Lake Erie to the Atlantic Ocean.

Eaton proclaimed this profile to be the first transverse view of rock strata across such an extensive district anywhere on Earth, based on "the particular examination of each individual rock." His profile contained an idealized cross-section

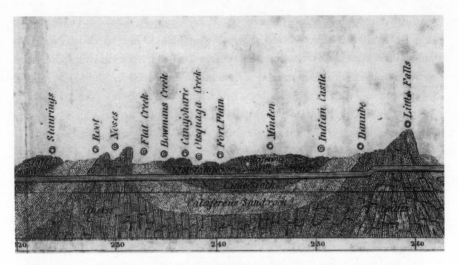

Figure 3 Detail from Amos Eaton, *Geological Profile Extending from the Atlantic to Lake Erie*, 1824. Eaton's audacious five-hundred-mile window into the arrangement of strata presented his best guesses regarding the underlying relationships among North American rock formations. In this small portion, he considered the saliferous slate exposure near Minden, New York, to be a localized bed of rock resting upon a slightly larger bed of metalliferous lime rock generally observed both to the east (Canajoharie) and to the west (Indian Castle). Eaton surmised that the adjacent calciferous sand rock provided the base layer for these nested beds, all of which resided atop a regionally universal layer of gneiss. Eaton inferred that the older and harder gneiss must have resisted the wear of ages (including at least one episode of catastrophic flooding) to continue as the most prominent topographical feature along this dramatic stretch of the Mohawk River valley (Root's Nose and Little Falls). Taken from Amos Eaton, *A Geological and Agricultural Survey of the Erie Canal* (Albany, NY: Packard and Van Benthyusen, 1824). The entire profile is available at http://memory.loc.gov/ cgi-bin/query/r?ammem/gmd:@field(NUMBER+@band(g3801c+ct000233)).

of the geological strata found in bedrock exposures stretching from Boston's harbor to Lake Erie. The profile exaggerates the vertical scale of elevations above sea level and illustrates how Eaton tried within the limits of a two-dimensional diagram to represent the contours and orientations of the strike faces (the angles of intersection between each distinct formation) and the dip slopes (the angles of inclination of the layers comprising each formation). Eaton's accompanying note of congratulations to Van Rensselaer reflected his pride in their combined accomplishment: "If your directions have been faithfully executed, you will have presented to the science more *elementary facts in a connected view*, than any other individual."[22]

Besides recording and organizing his scattered observations, Eaton had collected a massive number of rock and fossil specimens. These specimens were a
valuable commodity in themselves and might have generated a substantial amount
of revenue had he been practicing scientific research under the conditions that
governed the behavior of his more mercenary contemporaries. Assured by his
patron's apparent readiness to sustain these research endeavors, Eaton advised
Van Rensselaer to make gifts of twenty-five complete sets of rock and mineral
samples to various celebrated mineral cabinets. These sets were to be distributed
strategically so as to garner the most prestige for Eaton and his benefactor; the
list of designated recipients included wealthy private collectors and public institutions within the United States, as well as scientific societies in London and
Paris that had previously accorded honors and recognition to David Hosack and
DeWitt Clinton.[23] Finally, to demonstrate to state legislators that the economic
benefits of geologizing extended far beyond commerce in specimens, Eaton had
also done his best to identify and promote promising sites for mining across New
York State.[24]

Amos Eaton had arrived as a leader in American science. His strenuously
packed calendar attested to his reputation. Every fall, he gave his natural history
courses to the medical students at Castleton. In winters, he delivered lectures at
the Troy Lyceum, despite the illness and hoarseness that consistently plagued
him in cold weather. Every spring, he lectured to the female students at Emma
Willard's Troy Academy, and he accepted invitations to lecture at other colleges.
For example, he gave a short course in botany at the West Point Military Academy in 1822. In 1823, he taught geology and zoology alongside Edward Hitchcock
at Hitchcock's "new college" in Amherst, Massachusetts.[25] Meanwhile, with the
Patroon's continued financial and intellectual support, Eaton pressed ahead with
his field research, brimming with confidence that a sequel to the first Erie Canal
report would be both feasible and desirable.

SUBTERRANEAN FORESTS

Barely breaking stride after the publication of his first Erie Canal survey report,
Eaton resumed his geological journal in the spring of 1824 with new instructions
from his patron: "[Van Rensselaer] wishes me to examine and describe the alluvial formations—caverns on Buckland's plan—method of culture—plants and
animals, &c."[26] This list signaled a major shift in Eaton's natural history research.
Now, rather than probing to reveal hidden mineral treasures and the deep structure of the terrain (disclosed by the pattern of bedrock stratigraphy), Eaton would

turn his gaze upon the mysteries of the surface debris, some of which had clearly been transported some distance.

In 1821, Oxford's professor of geology, Rev. William Buckland, had stunned the scientific world with his revelations about the great Flood in his analysis of the large mammal bones he discovered at Kirkdale cave in Yorkshire, England.[27] Within the next couple of years, diluvial geology was elevated from a collection of vague scriptural allusions to a systematic framework for explaining flood-induced phenomena. According to biographer Nicolaas Rupke, Buckland's 1823 *Reliquiae Diluvianae* provided the dominant paradigm for geological practice in England until after Lyell's *Principles of Geology* appeared in the early 1830s.[28] Americans were well informed about Buckland's treatise by Hitchcock's comprehensive review, which Silliman published in 1824.[29] The English subtitle, "observations on the organic remains contained in caves, fissures, and diluvial gravel, and on other geological phenomena, attesting to the action of an universal deluge," hints at Buckland's method and his thesis. Eaton, usually reluctant to acknowledge foreign authority in scientific matters, gave the Englishman high praise for leading the way for investigators of the Deluge to produce valuable geological research: "Since [1820] Buckland, aided by the veteran Cuvier, has commanded the whole geological phalanx to leave for awhile the deep abodes of rocks, and to examine 'the open caverns and the furrowed earth.' He has led out before us, from the cave of Kirkdale, the antediluvial mastodon, chased and gnawed by hundreds of hyenas." Eaton went on to credit Buckland with a critical methodological leap: "The deluge no longer rests on the authority of written evidence. He points to records as durable as the earth, and far less changeable. The study of organized beings [i.e., fossils] has become the most essential qualification for the study of geology; for their relics are the most sure guides to truth."[30]

By 1824, Eaton and his patron Van Rensselaer were keenly aware of an opportunity to make an important scientific discovery. Aside from Schoolcraft's speculative report of antediluvial "human footprints" encased in lime rock quarried near St. Louis, no one had yet found evidence in North America of the human remains which could clearly identify the great flood as Noah's. Widespread discoveries of mammoth bones and the tremendous bounty of boulders indicated to Eaton that New York State should be a prime hunting ground for the crucial proof. All Eaton had to do was find the right cave or to be notified when canal workers or building foundation diggers disturbed the right pile of sand and gravel. The most enticing question in contemporary historical geology beckoned.

Embarking upon his fifth canal tour, Eaton was immediately induced to invent his own version of the diluvial theory, thanks to the hospitality of Jeremiah Brainard, a Rome, New York, contractor. Brainard was locally famous for inventing a tool to replace the traditional cumbersome wheelbarrow when transporting and unloading dirt. His lightweight alternative was "a single flexible board" bent "into a semi-circular shape."[31] Eaton happened to be staying at the inventor's home during a stopover on his way westward that spring, when he was astounded to learn of a sunken forest that Brainard had encountered while digging a well for his canal-side home: "Mr. Brainard tells us, that in digging his well (which is yet not completed) in the summer of 1822, at the depth of twenty eight feet he came to a red clay. On this clay is a layer of wood ten feet thick, over which is gravel. Immediately upon the red clay he found the leaves of the hemlock tree (*pinus canadensis*) yet of a green colour. Also the strobiles in full perfection. The water at this depth is not salt, nor in any way different from the swamp water of the surface." Eaton brought his expertise in North American botany to bear. "It must be remembered that the pines on the surface (28 feet above) are some of the longest and oldest trees of the forest. Consequently this alluvial deposit [*sic*] must have been made many centuries ago, though the leaves, strobiles, &c are still so perfect."[32]

Since Eaton was also engaged to present a series of public lectures in Rome and Utica throughout the next six weeks (this fifteen-mile interval of the Erie Canal had already been opened to commuter traffic), he made careful plans for how to investigate the subterranean forest during his prolonged stay.[33] On 16 June he traveled to the swamp to examine the well and to carry out other excavations into what he still considered to be alluvial debris. An extended entry from his journal records the full flight of his ambitious empirical research plans and his excitement at the theoretical implications:

> I engage Mr. Brainard to bore or dig several holes in this swamp at 50 cents per foot. I intend to give particular attention to this subject. I have good reason to suspect the existence of a subterranean forest in this swamp at the depth of from 15 to 30 feet.
>
> I must make a thorough examination for similar deposits all along the canal line. At about the same depth and under an alluvial formation very much like this, we find similar timber in Troy. Cannot I demonstrate from an examination of facts, that there are numerous forests in various parts of our district, which were buried at the same time?
>
> The trees which I have examined have generally, perhaps always, been the hemlock tree (*pinus canadensis*) at Troy, this place and elsewhere. Perhaps

the hemlock tree was most abundant in North America at a very ancient date. Perhaps a general deluge overwhelmed this country, and washed the alluvial deposites [sic] from the elevated parts into the valleys, basins and lakes. Where there happened to be a forest at the foot of a hill, as at the foot of this graywacke hill of Rome, or at the foot of the graywacke ridge east of Troy, the forest was broken down and covered to a greater or less depth. I must be particular to ascertain whether the timber is the same and of about the same state of preservation. Its preservation however will differ if of equal age, according to the nature of the alluvion.

I must see whether the alluvion was washed in one direction more than in another. Perhaps I may thus trace the direction of the water. I must be very particular about the vegetables of the subterranean forest.

I must be particular about the bivalve molluscas, which were found alive in the solid earth while digging the canal here—also their prolific increase along the canal since water was let into it. —And the minivalves between Limestone Creek and Salina; and their increase along the water of the canal, their adhering to the walls of the docks &c.[34]

Like Buckland, Eaton sought to mobilize all fields of scientific inquiry. He immediately entertained paleobotanical and conchological assessments of the materials he hoped to find by drilling holes into the ground. Eaton no longer depended on traditional Wernerian mineralogy alone. To sustain his new research program into earth history, Eaton was willing to try whatever tools he could imagine.

Two days later, his brain spinning with anticipation, Eaton asked Mr. Brainard to dig further down into his original twenty-eight-foot hole, "that I may examine this alluvial deposite [sic] still deeper."[35] It is important to note that Eaton's terminology did not yet embrace Buckland's diluvial geological framework. Carefully distinguishing between "alluvion" (soil moved to a spot by observable water sources) and "diluvion" (rocks and soil whose present situation could only be attributed to flooding of a much greater force and extent) would be a critical step in the refinement of Eaton's own theoretical scheme. From his public lecture notes, it is clear that he was working feverishly to develop a theory that would account for the buried hemlock trees, even as he anticipated a vast expansion of experimental data: "Lecture at Utica this morning on petrifactions. I make this remark, which seems to be proved by extensive observation. That *we have not as many organic relics on this continent as on the eastern continent in the same geological formations.*" He considered that this discrepancy between the Old and New Worlds might be a significant clue to the distinctness of their experiences of geological time. "If this proves to be correct on a more extensive course of observations, what is the cause? Will it not go towards proving, that the eastern continent was first inhabited in antediluvian as well as in postdiluvian times?" Or, the discrepancy might just as well be a symptom of inadequate American geological fieldwork. "We must not be too hasty, however, in deciding that we have fewer species of relics in the same formation. We have no chalk formation here, which contains the greatest number of petrifactions in Europe. But we have caverns, and deep secondary alluvial deposites [sic]. These must be searched for relics." In any case, Eaton's research program promised answers to some of the most perplexing issues then facing geology. He listed the driving questions as follows:

Was the ancient continent in the East warmest before the deluge?
Was this caused by greater internal heat?
Was the great heat any way connected with the more frequent primitive ranges in the East?
Is not the chalk formation in some way dependant [sic] on the ancient peopling of the eastern continent? Does not all those recent formations on that continent, which are not found on this, depend on the vast population for thousands of years? Or why does not our vast and varied territory somewhere present small patches at least of those formations which are so very common on the eastern continent? We find all the primitive, transition and oldest secondary rocks. If the same causes have always operated here, why have they not produced some of their results?[36]

In short, Eaton wondered how and why the geology of North America might deviate from that of Europe and Asia. By articulating this agenda, he crystallized a set of assumptions and hypotheses that would serve him well in his attempts to set himself apart from other North American natural historians and liberate North American geology from its European models. By doubting the universality of rock formations around the world, Eaton reaffirmed the importance of American fieldwork and escalated the primacy of New York as a site of geological investigation. In addition, by speculating upon environmental conditions during "antediluvian times," Eaton also opened the way for Americans to substantiate the Flood without necessarily having to produce the remains of one of Noah's drowned contemporaries.

Eaton became suddenly ill again in mid-July. He retreated from his fieldwork to recuperate in Troy for a week and a half but then returned to western New York before he was fully recovered. On 4 August, he boarded a steamboat that would carry him the length of Cayuga Lake. With the investigation of Rome's subterranean forest still very fresh in his mind, he pondered the dramatic evidence of multiple phases of debris deposition, apparently regardless of present topography: "The distinction between primary and secondary alluvion is more and more clear and interesting to me. It seems to be founded in nature, and of much importance. It seems as if the primary alluvion has been deposited from water which stood over the earth at a vast depth. Hence its layers are horizontal or undulatory, having no refference [sic] to the present slopes of hills or hollows." Eaton began to devise complicated hypotheses to explain Cayuga's dramatically sculpted shores: "Afterwards channels and deep hollows seem to have been excavated—many by a cause which has ceased to operate. Especially those excavations which were made in places surrounded by rocky brims, over which no rivers could have carried the alluvial matter as they could possibly flow in the present state of things. Vast and rapid currents in the ocean, under the pressure of great depth of water, seem to be the only assignable cause."[37]

These observations led Eaton to invoke some rudimentary physics principles as he tried to carry his geological thought experiments toward some conclusions: "As the pressure increases with the depth of the ocean, if set in motion it could not be resisted by the most compact alluvial substances, or whose constituent pebbles were of the largest kind. Boulders of many tons weight would be borne on before it, and wafted over the highest rocks in the direction of the current, if their elevated parts were still at considerable depth." Eaton now directly confronted the mysteries of the debris. For those who know subsequent history of geological theory, it is difficult to resist seeing his speculations as a succinct summary of the case that would later be made on behalf of the glacial hypothesis: "The extensive distribution of the durable kind of primitive rocks, several hundred miles from their original places, proves conclusively, that a cause like this has operated long and powerfully. The vast swamp of the canal district may have been scooped out in this way. Even the softer kind of rocks would even [sic] yield to this force, when partially disintegrated, and thus aided in scooping out the basins of Lake Erie, Lake Ontario, Lake Champlain, &c." Tying his newest Finger Lakes evidence back into the subterranean forest problem, Eaton concluded his theoretical speculations with a final bow to that evidence: "After the subsidence of the water . . . the secondary alluvion was washed down into those excavations, carrying down hemlock trees and other vegetables perhaps. The hemlock tree is most common, proving that it greatly abounded in an early period. Sometimes (or perhaps at one time only) whole forests seem to have been overwhelmed by a general deluge, and thus to have formed the subterranean forest at Rome[,] the great swamp, at Troy and at other places."[38]

This journal entry from early August 1824 marks the first full conceptualization of Eaton's diluvial solution to the variety of mysterious phenomena he had observed. The distinction between "primary" and "secondary alluvial" debris would provide a key to his revised nomenclature and would lead to a refined description of what the deluge could have been like.

REVERIE AT ROOT'S NOSE

Returning eastward along the Mohawk River, Eaton was still trying to find good cave evidence as his fifth canal tour was wrapping up. Near Spraker's Basin, the river bends and passes through a narrow gap where hills on either side of the river rise up more than four hundred feet, even though the valley is much wider both above and below this point. Upon the hill abutting the south side of the gap, Samuel Latham Mitchill had recently discovered a narrow entrance to an elevated cavern that descended steeply into the Earth's interior for several hundred feet. Naturally, Eaton and his entourage scaled Root's Nose to investigate Mitchell's Cave,[39] hoping to find bones like those Buckland had analyzed. Leaving one of his sons and three other young assistants inside to map the cave's structure and survey its contents, Eaton returned to the surface for a rest. Perched there outside the cavern, Eaton pondered his lot as a storm began to brew. Describing this occasion, Eaton's usually dry and businesslike prose gives way to a reverie full of feeling and allusions. The tumultuous weather set the tone. Reminiscent of the Knickerbocker spoof, Eaton's journal entry gives voice to an unsuspected Romantic impulse: "the thunders roar tremendously." Like those poetic travelers Eaton usually abjured, his own spirit seems to have experienced a sublime uplift amid the tumult. His journal of field notes would never again replicate this momentary rapture; just this once he expressed his wonder at nature's size and power and his gratitude at the magnanimity of the person who had made his career as a scientist practicable. Here is the complete journal entry (the quote that closes this passage should be familiar from the epigraph to chapter 3):

> We find the cavern to be an extensive east and west fissure in the calciferous sand rock. In the fashionable language of puffing geologists, this fissure "was caused by a tremendous convulsion of Nature." But to those plodding examiners of rocks, whose poetic faculties are not well developed, it appears to be a very extensive crack in the rock governed, however, by the common cause as fissures operating on an uncommonly broad and deep mass of rock. . . .
>
> It is now one o'clock P.M. I sit here at the mouth of the cavern, waiting for messages from the subterranean travellers. A shower is rising and the thun-

ders roar tremendously. A violent wind begins to shake the forest trees which surround me. I have a little fire burning before me for lighting extinguished candles, which are occasionally sent up from the interiors of the earth. I have just peeled the half rotted bark from a pine stump over the cave's mouth, to cover the cloathes [sic] and books of my assistants.

We are highly animated with our investigations, and feel a pride in overcoming difficulties and in the prospect of being able to present interesting results to the learned and ingenuous. But who will remember us or our labors for half a century? Which is but two years longer than I have already lived. Who will remember the beneficent Stephen Van Rensselaer? The myriads of ambitious mortals who have preceeded [sic] us are forgotten. So we of the present generation who are wearing down our strength in climbing precipices and descending caverns, cannot hope to be remembered but a few years. Why should Van Rensselaer send us here at great expense, when he too is so soon to be forgotten? It must be that he has ungovernable propensities to do good, which are as unmanageable as the thirst of the drunkard.[40]

Eaton had good reason to be thankful. His research program was now as robust and promising as it could conceivably be, and yet he still craved something more. Feeling his age, he worried that he might soon succumb to one or another of his periodic illnesses without having established a more permanent legacy in American science. It occurred to him to reconsider whether one's published research record was the only measure or means of creating that legacy. From the end of his fifth canal tour in August 1824 until the spring of 1826, Eaton would focus much more of his attention on the establishment of a new college rather than on the immediate continuation of his geological research program.[41]

Soon after his return from Root's Nose, Eaton proposed to his patron the idea of a "Rensselaer Institute" in Troy. The Patroon was favorably disposed from the outset, viewing the institute, in historian Michele Aldrich's words, "as an internal improvement project similar to the Erie Canal."[42] Eaton's skills as an instructor were now beyond question, and Clinton and Van Rensselaer were encouraging sons of their own to enter Eaton's scientific tutelage.[43] The Patroon ultimately obliged Eaton by providing sufficient financial resources to purchase apparatus and a building called the Old Bank Place in the north end of Troy. Eaton was officially appointed in November 1824 to serve as the new institute's senior professor. He would be responsible for instruction in chemistry, experimental philosophy, geology, land surveying, and public law. Lewis Beck was appointed to the junior professorship and would lecture on mineralogy, botany, zoology, and "the social duties peculiar to farmers and mechanics."[44] Van Rensselaer's agreement to pay all the initial costs and half the operating expenses made possible a school whose mission was to spread knowledge of science among New York's farmers and tradesmen. Ultimately, Troy would become home to the first civilian school of engineering sciences in the United States.[45] The consequences for New York State and for American science were profound. Eaton's promotion to a stable, full-time professorship meant that a new generation of professional men would receive training from a man who understood and practiced a hands-on approach to the field sciences. Every student who attended the Rensselaer School during its first dozen years was fully inculcated in Eaton's vision of science instruction as being both a practically valuable experience and a collective endeavor of people fully engaged in the development and articulation of an improved theoretical understanding of the historical and dynamic agencies that had shaped all the features to be found in New York's complicated physical geography.

CHAPTER SIX

Empire State Exports

Whoever is "first in the field" of natural science, has an exclusive right *to give names*. His successors should either adopt his names, or give them as synonyms and equivalents. This is essential to the very being of science. But English and French geologists have introduced new names, not adopted in Germany; because new discoveries have made them necessary. I have done the same thing in America, and for the same reasons.

— *Amos Eaton, 1830*

Amos Eaton made no research excursions along the Erie Canal route in 1825, but an extraordinary parade of public figures did travel from Buffalo to New York City in the celebration that officially opened the completed canal to navigation. As Eaton had foreseen, the new era brought a massive influx of migrants heading westward, as well as a growing number of tourists who wished to see the wondrous natural sights of New York State. By all accounts, the "Wedding of the Waters" on 26 October 1825, was a glorious and noisy affair. A flotilla of boats loaded with symbolic cargo convened at Buffalo. At the head of the procession, the *Seneca Chief* carried kegs filled with Lake Erie water, potash from Detroit and Buffalo, flour and butter from Michigan and Ohio farms, a canoe made by American Indians from the shores of Lake Superior, and bundles of cut red cedar and bird's-eye maple wood for making boxes to hold medals to commemorate the event. A harvest of unprecedented prosperity was to be reaped by this symbolic unification of the nation. But amid the hyperbole, some legitimate accolades were showered upon those talented social engineers—DeWitt Clinton first among them—who had coaxed New York science, politics, and technology into fertile embrace.

After crowds heard Governor Clinton and his fellow canal visionary Jesse Hawley speak in Buffalo, the procession commenced. Behind the *Seneca Chief* came the *Superior*, followed by the *Commodore Perry* (a freight boat named after the War of 1812 hero of the Battle of Lake Erie), the *Buffalo*, and finally a boat "fitted out to portray the wilderness as it was before the white man came." This last vessel, dubbed *Noah's Ark*, was a barge assembled to serve on this one occasion as a parade float; it carried two young Seneca Indian boys, two eagles, a bear, two deer, a beaver, a cage of wild birds, and a tank of fish.[1] News of the flotilla's departure raced ahead of the procession using the fastest communications technology then available: bursts of cannon fire were relayed from one town to the next along the entire route. Eighty-one minutes after the first cannon had been fired in Buffalo, residents in New York City heard a rumbling blast. City leaders immediately returned the salute that ushered in New York's new preeminence as a gateway to the center of the continent. Nine days later, on 4 November 1825, the ocean-bound boats themselves arrived from Buffalo, having been joined by forty-two other vessels. The *Seneca Chief* ceremoniously disgorged its store of Lake Erie water into the Atlantic Ocean, thus consummating the marriage of Great Lakes and seaboard commerce that Clinton had worked so long and hard to realize.

Through a combination of political will, persistent leadership, and the fortuitous geological circumstances of its terrain, topography, water resources, and a recognizable native limestone supply, New York had managed to cobble together a structure that would invite both a mass influx of coastal American people into the interior of the continent and an efflux of natural resources and agricultural produce, more or less fulfilling the promise of continental transformation depicted by the Wedding of the Waters. Generations of American historians have produced groundbreaking works in their turn, each tying major societal shifts to the opening of the Erie Canal, including the emergence and spread of novel religious cults and utopian community experiments across what became known as the "burned-over" district and the galvanizing of political and market revolutions in the Jacksonian era.[2] Fewer historians have traced the far less grandiose but immensely practical and intellectually significant collateral effects of science's essential complicity in DeWitt Clinton's program of internal improvements. These unsung "exports" of 1820s New York include Eaton's diluvial theory of surficial geological deposits, the educational philosophy and apparatus Eaton developed to turn out trained professional scientists, and Clinton's idea that science could be utilized to transform virtually any landscape into a profitable canal. Moreover, New York's exposure of such a lengthy slice into Earth's history translated into an original system of geological nomenclature and a major step forward in geological mapping techniques.

THE GREAT DILUVIAL TROUGH

Far from the bombast of Governor Clinton's shining moment of triumph, Eaton quietly deliberated about the spectacular evidence he was finding on behalf of past inundation of the entire state. In all likelihood, the image of *Noah's Ark* cruising past Troy had no direct influence on Eaton's thoughts, yet he was preparing to christen the territory through which the canal passed "the Great Diluvial Trough." Within the next thirteen months, Eaton assembled a comprehensive array of evidence for the deluge, evidence that modern geologists would instantly recognize as archetypal glacial debris. Innocent of any ice age theory, Eaton had nonetheless come startlingly close to guessing that ice might have produced phenomena comparable to those he attributed to the agency of water. If only he had been reminded of his own observation of that "singular deposit of gravel" by the sloop locks of West Troy four years earlier, as he now struggled to explain how a flood could have produced the drumlins he found along the Mohawk valley and beyond. Instead of uniting the variety of fantastic glacial phenomena into a comprehensive documentation of planetary climate change, Eaton delineated, classified, and injected all these stray pieces of evidence into his emerging theory of diluvial actions.

To account for the variety of debris types and to handle the problems posed if one assumed that it had all been deposited in a single moment, Eaton proposed two categories of diluvion: the deluged (or chaotic) type and the quiescent (or mantling) type. Eaton presented these phenomena in a letter to Benjamin Silliman dated 26 November 1826: "I find a diluvial trough extending from Little Falls, along the Erie Canal, one hundred and sixty miles . . . The insulated remains of the stratified (antediluvial) deposites [sic], present the marly clay, bagshot sand and crag, beautifully crowned with almost snow-white shell marble [sic], a fine yellowish soil, and vegetable mould, or peat." He went on to explain the interpretive challenges these jumbled substances presented to the historical geologist: "I may add, that nothing is more manifest, than that these deposites could not have been made by any existing cause. Seventy miles of this region is occupied by the summit level of the canal. The surrounding country is but a few feet higher, and all the water naturally flows into Lake Ontario, or through the stupendous chasm at Little Falls."[3]

Eaton closed his landmark report by suggesting how diluvial actions happening in a certain sequence might have created such a convoluted result: "It is, as it would have been, the whole [of the 160-mile diluvial trough] having been filled to its present level with marly clay, covered with bagshot sand and crag, generally overspread with a layer of shell-marl, had it then been cut up, by a strong current running from Little Falls westerly, into islands, ridges, embankments, &c.; and after these channels were thus made, had they been filled with a confused mass of gravel, sand, clay, trees, leaves, fresh-water shells, &c. Whether the appearances originated in this manner, or in any other way, such is the present aspect."[4]

Before he dared to share his provocative ideas with his colleagues and the general public, Eaton privately formulated a highly detailed explanation of the putative flood's effects. He confided and organized his ideas in a monumental entry to his geological journal. "The *deluged* or *chaotic* diluvion was deposited while the deluge was at its greatest strength. The *quiescent* or *mantling* diluvion was deposited when the violent action of the water had nearly subsided. It was held suspended in the waters, like a similar suspension of earthy matter in our rivers at this day in time of a freshet." Worrying about the global dimensions of a sufficiently generalized flood event, Eaton wove his speculations into the fabric of his theory-laden description. "But the water being several miles in depth (which was necessary for covering the Andes) a mantle of from 6 *to* 24 *inches* was deposited. Some of the lowest part settled down while the waters were considerably agitated. This is coarser and contains large pebbles. As the waters ran gradually off, after the cause which raised them had ceased to act, the deposits became finer, until at length it became of the fineness which we now discover in the *ancient undisturbed forests*. This accounts for the presence of a fine soil spread over the hard-pan (crag) of our western woods, which is totally unlike any of the earths in connexion [*sic*] with it, or within many miles of it." Through his extended ruminations, a straightforward physics problem (decoding the sequential precipitation of different solids out of a turbid solution) appears to have preoccupied the self-taught natural scientist: "In many places the *deluged* diluvion would be formed, while the deluge was at its greatest strength; and the *quiescent* diluvion be deposited upon it after the violence of the water had subsided. In such localities the *'mantling'* diluvion would differ materially from the *chaotic* deposits, like the vast heterogeneous masses in the great diluvial trough. Such is ~~naturally~~ really the appearance of those localities."[5]

Eaton was well aware of the fact that the state's inhabitants were bound to disturb and displace evidence as settlement and development progressed. Therefore, he felt it necessary to plan carefully for how field research in geology should proceed in order to document this dramatic episode from Earth's history: "While examining the diluvial mantle, we must guard against mistakes. In cultivated fields, this mantle is greatly disturbed by the plough, &c. And even in the most ancient forests, there are numerous localities, where great changes have been produced by freshets. At and near the bottoms of hills, and even on moderate slopes, new portions of earth have been washed upon the diluvial coat." To find the most pristine circumstances of predisturbance geological evidence, Eaton advised his disciples to venture into ancient forests situated upon hardpan soils on extended plains of a high relative altitude. Promising locales in New York included the towns of Duanesburg and Blenheim and the elevated recesses of the Catskill Mountains. Of the soil in these remote districts, Eaton observed: "Here the hard-pan is covered by a few inches of fine earth, generally of a yellowish colour, differing on the whole widely in its character from any other layer of earthy matter in the district; though it is generally intermixed with them. It appears to have been transported from some distant region, as its constituents cannot be the ruins of any neighboring rock, nor the broken up materials of the antidiluvion of that vicinity. . . . [At the cavern at Root's Nose] we found the same kind of diluvion at the distance of 400 feet from its mouth, that overlays the hard-pan of the adjoining high country to the south of the cavern. But it might have washed in from the diluvial mantle about the mouth." Eaton concluded his practice walk-through of New York's supposed experience of the great flood with the following summary: "From these facts it seems that we may safely infer, that during the deluge there was a time when the waters moved with the greatest conceivable degree of force. Their impetuous currents ploughed deep channels in the antidiluvion [sic], and filled these again as their directions and velocities varied, with the heterogeneous rocks which are now to be seen, of very great and variable thickness. And that towards the end of the deluge the agent which gave motion to the waters and caused those stupendous ravages upon the surface of the earth, was withdrawn; leaving the waters to glide away, seeking their equilibrium according to the laws of gravitation."[6]

Eaton's preparations to announce the theory to the American scientific community included consulting with his patron. On 10 November 1826, less than two weeks before sending Silliman his summary report, Eaton met with Stephen Van Rensselaer, who urged him to "compleat and publish the 2 part of the Canal Survey, as soon as possible." Eaton considered his diluvial theory to be the centerpiece of the work. On 13 November, he wrote in his journal: "*I now intend to propose the following new hypothesis. That[,]* near the last days of the deluge, when the strength and violence of the waters had abated, and they were subsiding by the common law of the equilibrium; the last and, consequently the finest sediment was deposited upon every formation which was then uppermost. That this sediment formed an *universal mantle*, enveloping the whole earth before

it was disturbed by natural or artificial agents; unless some of the highest mountains should be found to be exceptions on account of their projection abouve
[sic] the waters before they became sufficiently quiescent for such a deposition."[7]

TRAINED PRACTICAL SCIENTISTS

By the time Eaton had conceived his systematic theoretical apparatus for discuss-
ing "diluvion," his experiment in practice-based pedagogy had already fundamentally changed the world of science education. Among the many remarkable
American scientists produced because of this scheme was Douglass Houghton.
Born in Troy but raised on the Chautauqua County farmlands of the Lake Erie
shore, Houghton was brought back to Troy to be educated according to Rensselaer's plan, attending lectures and acquiring the mechanical and scientific skill
to aid farmers living in his distant corner of New York State. He graduated from
the Rensselaer Institute in 1829 at the age of twenty-one, but instead of returning to serve the needs of the citizens of his native county, he remained in Troy
for another year as one of Eaton's adjunct instructors. Upon the satisfactory completion of those responsibilities, Houghton followed Henry Rowe Schoolcraft's
example by seeking more lucrative and interesting scientific opportunities out
west. He took his first job as a lecturer of science at the Detroit Military Post. He
proceeded to organize the Michigan Geological Survey, become the mayor of
Detroit, and eventually was hired as the first professor of geology at the University
of Michigan. Tragically, Houghton drowned at the age of thirty-six on an expedition to investigate reports of remarkable copper finds at Lake Superior.[8]

How had the Rensselaer School transformed a farm lad like Houghton
into such a valuable emigré from his home state? By establishing a laboratory-
centered curriculum for the explicit purpose of practical scientific education,
Eaton's influence on the training of North America's first generation of professional earth scientists rivaled that of his European contemporary, the German
agricultural chemist Justus von Liebig. Liebig had been invited by Alexander
von Humboldt to institute a comparable pedagogical experiment at Giessen in
1825.[9] Like Houghton before him, Ebenezer Horsford was a promising youth from
the western part of New York State who was sent to Troy to be educated in the
natural sciences. Horsford studied under both Eaton (graduating at the age of 20
in 1838) and Liebig (working in Germany from 1844 to 1846) before bringing the
modern laboratory science curriculum to Harvard College. Horsford was Harvard's Rumford Professor of Chemistry from 1847 until 1863, when he retired to
devote his full attention to the thriving business venture he had founded in Providence, Rhode Island: the Rumford Chemical Works, whose leading product was

"Horsford's Self-Raising Bread Preparation," a commercial baking powder whose invention brought Horsford wealth and fame. At the Rensselaer Polytechnic Institute's semicentennial celebration in 1875, Horsford spoke about how similar Eaton's laboratory instructional methods at Troy were to Liebig's at Giessen.[10]

The Rensselaerian plan of education, which brought such notoriety to Eaton as a teacher, was built upon the premise that learning requires the active involvement of the learner. To become a scientist, Eaton had first studied leading European texts on his own and then attended Silliman's lectures at Yale. Most important, he had covered thousands of miles of terrain on foot, studying evidence from nature firsthand and in situ. His experience gathering ideas and materials had always served the dual purpose of informing his scientific writing and enlivening his attempts to explain and demonstrate the wonders of natural history in lectures before the general public. In his teaching, he therefore strove to recapitulate and concentrate the positive features of his own education. He sought to provide his students with the benefits of a self-guided reach for knowledge, free of the accidents and suffering he had endured in his idiosyncratic career. Under their professor's direct supervision, students at the Rensselaer School underwent an intensive regime of laboratory work, lecture, and recitation. They were required to "teach" to their peers any materials with which they were becoming acquainted. For evaluation purposes, outside experts and leading public figures were invited to attend the students' lectures. Eaton was just as proud of the results of his educational innovations as he tended to be about his scientific discoveries. The method of teaching by lectures and having students rotate record-keeping responsibility as "officer of the day" resembled elements of the European systems of general pedagogy espoused by Joseph Lancaster and Emanuel von Fellenberg, then both much in vogue. To contemporaries teaching at other colleges in the northern United States who expressed doubts that Eaton's method of practical instruction was really any different or new, Eaton retorted, "The unwillingness to admit the *possibility* of an American improvement in the course of education which generally prevails, and the universal homage paid to everything European, has caused much effort to trace the Rensselaerian plan to some supposed shade of it on the other side of the Atlantic."[11]

The most profoundly inventive aspect of Eaton's pedagogic experiment was the annual nature expedition, which the senior professor inaugurated with the first class in 1826 in his program to prepare scientists who would be competent at performing public lectures.[12] In keeping with his goal of providing a maximum of hands-on experience to his students, he chartered the canal boat *Marquis de La Fayette* to serve as a "floating school." Thus, his maiden instructional excursion coincided with his sixth research tour of the Erie Canal route. His journal entry for 2 May 1826 recorded everyone in attendance: "Besides the captain and three hands, the following young gentlemen embarked as members of the expedition. *Rensselaer Students*[:] Stilman E. Arms, Albert Danker, Hezekiah H. Eaton, T. Dwight Eaton, Richard H. Hale, Oscar Hanks, Addison Hulbert, Saml. C. Jackson, P. C. W. T. Mc-Manus, Wm. S. Pelton, Bennett F. Root, James M. Trimble, Chas. Weston, Richard H. Williams. *Persons not of Rensselaer School*[:] Geo. W. Clinton, James Eights, Asa Fitch, Jr., Ebenezer L. Cady, Joseph Henry & James Talmadge Hildrith."[13] Of special interest here is the inclusion of those "persons not of Rensselaer School," most of whom were brought along to serve as research assistants and to model the practice of lectureship for the students. Two of Eaton's sons were among the Rensselaer students, and two of the fellow travelers had particularly distinguished futures ahead of them: James Eights would serve as geologist, zoologist, and botanist on Captain Edmund Fanning's Antarctic expedition in 1830,[14] and Joseph Henry would make world-renowned discoveries in the physics of electricity. Between 1 May and 10 June 1826, this group covered seven hundred miles, venturing from Troy to Buffalo and back, less than a year after the Erie Canal had become fully operational.

Aboard Eaton's floating school, every student was required to maintain his own geological journal, and the responsibility of head assistant rotated daily so that everyone had an opportunity to demonstrate his knowledge. Eaton wanted students to learn science through active personal engagement with phenomena and through the challenge of instantly articulating their thoughts via impromptu lectures to their peers. The field trip was intended to help students identify specimens in situ, that is, in their own surroundings far from the artificial organizing principles of mineral cabinets and fossil collections. These students were encouraged to speculate intelligently about the geological history of the terrain through which they were traveling. Eaton's journal mentions didactic discussions about how to identify whether a rock specimen belongs to a bed or a stratum, how to think in terms of stratigraphic superposition, and how to apply the action of "existing causes" to explain the emplacement of the phenomena that the students encountered. Joseph Henry's journal attests independently that the rotating of duties among the students did not amount to permissiveness. Eaton maintained a strict atmosphere of disciplined research aboard the boat.[15]

At the same time, Eaton utilized his students to vigorously test his own emerging diluvial theory. He tried to demonstrate its consequences in the field and encouraged his followers to question and challenge every speculative assertion. Now primarily engaged as a teacher, Eaton thoughtfully reformulated his ideas about

how geological investigation ought to be conducted. The field trip provided idealized circumstances for a consciously collective scientific effort, with the students
independently examining every locality. In particular, they were being taught
that every place can be described in terms of a twofold set of features: the bedrock composition and the covering, which Eaton regarded as either "anti-" [sic],
"post-," or simply "diluvial." As the group discussions helped Eaton solidify his
own incipient theory (this trip preceded publication of his diluvial ideas by half a
year), he ventured stronger statements about the insufficiency of present causes
to account for the debris. The "great diluvial trough" began to be presented as a
firm notion, not just a new and suggestive turn of phrase: "At Oriskany I explain
the *diluvial trough*, or great swamp. Then that the diluvial deposite [sic] could
not have been washed here by any existing cause. There is no place from which
water could flow to produce this result, unless the quantity of water was increased
beyound [sic] the power of any *known cause*."[16] This claim would survive to become a central piece of the argument Eaton would make in the letter he wrote to
Silliman six months later. Eaton may have had such a high degree of confidence
about what was in fact becoming a risky kind of scientific claim (to rely on the
past action of causes that are no longer in existence) because he had tested his
claims against the rapidly maturing judgments of his two dozen young companions. The Rensselaerian method of training was so deeply imbued with the obligation to cultivate one's own sense of practical mastery in natural history that
Eaton invested some faith in his student's conclusions. These boys had been
given a standardized set of intellectual tools, to be sure. But Eaton's insistence
that they examine his observations and dispute his claims before either corroborating his conclusions or coming to different ones of their own imagining served
ultimately to validate and reinforce his own research results.

The close company of motivated young men, though fruitful for the refinement of scientific ideas, proved ineffective as a way for Eaton to ward off the debil-
ities of his own advancing age. After two weeks of constant travel and performance
as researcher and teacher, Eaton succumbed to a "violent fainting fit" at the
Genesee River. He was attended by doctors in Rochester, but they could not prevail upon him to delay or desist from accompanying the tour westward. Reports
of a mammoth or a mastodon tooth at Sandy Creek lured the group on despite
the approaching summer heat and humidity. Regardless of what beast it might
be from, Eaton determined the tooth had belonged to a "post-diluvial animal,"
since it had been found in a peat bog with "black loam under it." To properly
place the tooth within the geological sequence, Eaton scrupulously noted the
English geologist William Conybeare's pronouncements on scientific method

but remarked that Conybeare's advice (to always begin stratigraphic mapping with the layer of marley clay "and go up or down") was a moot issue in western New York, since the marley clay there always appeared lowest. In an entry written a few days before his fainting spell, Eaton had proposed that North American diluvial deposits must be different from those reported in Europe, simpler in the sense that marley clay (like Conybeare's London clay) and Bagshot sand were to be found everywhere, whereas plastic clay was found only in small quantities in New York and never as a distinct stratum.[17]

It is easy to see why the field trips became a popular tradition among Rensselaer students. They provided every student with practical experience in unfamiliar terrain, while holding the possibility of witnessing a serendipitous discovery in the company of experts. Unsurprisingly, when New York State mounted its natural history survey a decade later, three men who had been trained by Amos Eaton to look at nature firsthand and to think for themselves (James Hall, Ebenezer Emmons, and John Torrey) were chosen to lead major portions of the state's ambitious research task. A highlight of that first field trip illustrates just how Eaton's approach enhanced the student's ability to partake of scientific discovery. Transylvania University professor Constantine Rafinesque, who was traveling east from Kentucky, asked if he could come aboard the boat for its return trip from Buffalo. The presence of the colorful and controversial naturalist triggered a burst of excitement back in Rochester. On 29 May, Eaton's son Hezekiah Hulbert Eaton caught and skinned a fish from the canal. Rafinesque caused an immediate stir by christening it *Ostrognathus chloripterus*, a species of fish he claimed had never before been observed.[18] Joseph Henry's journal wryly noted Rafinesque's infamous reputation for displaying a "proness [sic] to make new genera & species,"[19] but Asa Fitch (who was later hired by Emmons to work on the New York Natural History Survey and would eventually serve as the first state entomologist) came away from his maiden journey aboard the floating school overcome by the thrill of pursuing field research in the presence of bold scientific thinkers like Eaton and Rafinesque. Fitch wrote in his diary: "I was only gone seven weeks and yet how much I have seen! How far I have been! What new ideas I have received! and how greatly my mind has been improved."[20]

Although Eaton's diluvial theory was ingenious, it would not survive the next decade's revolutionary developments in geological theories and practices. Nevertheless, the audacious natural historian paved the way for Americans to declare independence in matters of nomenclature and modes of practical instruction in science. His constant impulse to flout authority from Europe and to instill a sense of confidence in American practical superiority and theoretical legitimacy was a significant inspiration to his contemporaries. By the end of the 1820s, Eaton had done virtually all that he could to deploy his patron's resources in the cause of advancing scientific knowledge. To carry any further the program of systematic investigation and dissemination of the state's natural history would require the more active sponsorship of the state government, and the resulting opportunities for professional scientific achievement would open up careers for many of the young men Eaton had trained.

CANAL FEVER

New York's remarkably successful canal construction meant that Albany would not be the only place where this new crop of trained practical scientists could ply their skills as natural historians and engineers. For better or worse, the promise of vast profitability that DeWitt Clinton had persuaded his fellow New Yorkers to cultivate had been heard, and the investment would soon be widely imitated. As early as 1821, Clinton's visionary leadership in the creation of a transportation infrastructure had inspired the former central New York lawyer and naturalist Caleb Atwater to campaign for a seat in the Ohio legislature on a platform advocating canals, improved roads, and public education. Prior to the opening of the Erie Canal, Ohio's solitary commercial transport route required traveling close to two thousand miles downstream along the Ohio and Mississippi rivers to New Orleans and the Gulf of Mexico. Niagara Falls effectively blocked easy access to the Atlantic Ocean via Lake Erie, and overland travel to the Atlantic seaboard was prohibitively expensive. Atwater estimated in 1820 that a land journey over the mountains to visit his colleagues in the New England and Mid-Atlantic cities would take him four months and cost $500![21]

Persistent efforts by Clinton's enemies (Martin Van Buren and the Bucktails) effectively frustrated the governor's political ambitions again in 1822. Clinton was so worn down that he refrained from seeking reelection even though a little more than half of the Erie Canal had been completed. In the spring of 1823, an enthusiastic Atwater urged Clinton to consider running again for the presidency of the United States. Henry Clay's campaign was already underway, but Atwater promised to divert Ohio support away from the Kentuckian should New York put Clinton forward in 1824. Atwater's letter characteristically blended political tactics and scientific matters: "If you are not a candidate, Mr. [John Quincy] Adams will be supported by us, though we want a friend to internal improvements. Clay's friends are setting up new presses here and buying up old ones. If New York wants a President *now* is her time to begin. A press is wanted here and if New York wishes the support of Indiana, Illinois & Ohio, *no time is to be lost*. I would start a paper myself, at Chillicothe, if I knew you really would be a candidate. We are tired of slaveholders for chief magistrates. I shall send your specimens and some books in the spring."[22]

While Clinton remained silent about his intentions, a convention held at Steubenville, Ohio, went ahead and nominated him for president in 1824. By that time, however, his sights were set on recapturing the New York governorship in time to witness the completion of the Erie Canal. Atwater was doubtless disappointed in January 1825, when he offered the following impassioned plea: "How long will New York continue in a spirit of selfishness, to keep *Clinton* employed in advancing *her* own individual interests and glory? He belongs to the *Nation* and to the *world*, where his talents would find a field of usefulness as extensive as his mind is vast and comprehensive. To New Yorkers we say, lay aside all selfish feelings and give your *Clinton* to the Nation and to the World."[23] Four years later, in a state of anguish after Clinton's death in February 1828, Atwater predicted terrible sectional consequences should General Andrew Jackson, whom he supported for the presidency, be governed by "unprincipled and daring" advisors committed to perpetuating another southern dynasty (i.e., John Calhoun).[24] Had Clinton survived, Jackson might have turned to him as his strongest supporter among northern Republicans and perhaps rewarded Clinton's loyalty by appointing him secretary of state. Instead, Van Buren maneuvered himself into becoming Jackson's eventual successor.

DeWitt Clinton never became a national leader for the United States, but in reality, on the issue of internal improvements, New York had never been stingy with "her Clinton." Historians characterize the incredible period following the completion of the Erie Canal as a time when the entire country was infected by "canal fever." While it is true that canal projects were taken up actively by almost every state in the union, it would be misleading to imagine that this all happened in the absence of a vigorous public debate about the wisdom of pursuing these expensive public works projects. Clinton's legacy in American history as the "prophet of internal improvement" depended not only on his own achievements in New York but also on the energy with which he urged other states to replicate the same feats. Between 1823 and 1828, the neighboring states of Ohio, New Jersey, and Connecticut consulted with Clinton and his engineers in the design, construction, and promotion of their canal building enterprises, and in every case he was a generous and optimistic supporter. In 1822, Ohio's governor Ethan Allen Brown asked his fellow governor to recommend someone capable of surveying a route for a proposed canal between the Ohio River and Lake Erie. Clinton named James Geddes. Although New York stood to lose one of its most competent engineers, Clinton trusted that David Thomas could finish the task of completing the Erie Canal's western section on his own, and so Geddes was given permission to accept his new position as chief engineer of the Ohio Canal.[25] Clinton publicly endorsed Ohio's plan the very next year.

> The advantages of [the projected canal between Lake Erie and the Ohio river] are so obvious, so striking, so numerous, and so extensive, that it is a work of supererogation to bring them into view. . . . when we consider that this canal will open a way to the great rivers that fall into the Mississippi, that it will be felt not only in the immense valley of that river, but as far west as the Rocky Mountains and the borders of Mexico; and that it will communicate with our great inland seas and their tributary rivers, with the ocean in various routes, and with the most productive regions of America,—there can [be] no question respecting the blessings that it will produce, the riches that it will create, and the energies that it will call into activity."[26]

On 4 July 1825, just a few months before the triumphal parade of boats across New York State, the newly reinstalled Governor Clinton joined Ohio governor Jeremiah Morrow to celebrate the beginning of work to build the Ohio and Erie Canal. Eight years to the day after Clinton had performed the same ceremony in his own state, he and Morrow turned over the first spadefuls of earth at Licking Summit, the highest point of the proposed route.[27] State owned and maintained but funded by wealthy investors (including a $1 million bond provided by the fur magnate John Jacob Astor), the Ohio and Erie would be 308 miles long, requiring a 793-foot rise from Lake Erie to Licking Summit and then a drop of 413 feet back down to the Ohio River.[28] Construction costs per mile were slightly lower than they had been in New York, partly because of the experience gained and the technological improvements developed by the Erie Canal's engineers. Additionally, having completed the main portions of the Erie, cadres of experienced workers flocked from New York to the new project.

Major canal endeavors were soon launched in New Jersey, too. Responding to a plea by New Jersey's canal commissioners in 1823, Clinton personally inspected the proposed route to link the Delaware and Passaic Rivers. He even brought along his chief engineer Benjamin Wright to evaluate construction cost projections. The logistics checked out well enough, and Clinton's report explicitly linked politics and economics to the blessings of cultivating scientific literacy: "The revenue arising from the canal will for ever supersede the necessity of taxation, and will form a vast fund applicable to other internal improvements, to the diffusion of the lights of science and to the dispensation of the blessings of education." Adding that canals in Great Britain had proven invariably lucrative where they provided access to coal mines, Clinton went on to predict: "The demands of the city of New-York, and the other cities and villages on the Hudson, the consumption of various parts of New-England, and the manufactories of New-Jersey, for this indispensable article will for ever increase, and for ever secure a great revenue from the canal." Finally, Clinton called particular attention to the commercial value of natural history specimens: "Add to this, the fossils and the metals before mentioned, the products of the forest and the field, and the fabrics of art, and there is no question but that this canal will enrich New-Jersey in her finances, as well as in other respects."[29]

While the Delaware-Passaic Canal was a publicly funded venture, Clinton became directly involved with a competing private venture as well. The Morris Canal Company had launched an ambitious scheme to cross the state of New Jersey with an artificial waterway between the Hudson and the Delaware Rivers. Cadwallader D. Colden, a fellow former mayor of New York City (from 1820 to 1821) and the namesake grandson of one of colonial New York's greatest scientific luminaries, was president of the company. In 1827 Colden invited Clinton back to see "the inclined planes at Rockaway, and to inspect the eastern division of the canal from the summit level to the Hudson, at the city of Jersey." Clinton obliged with a message calibrated to attract fresh infusions of cash from new investors: "The prospects of abundant remuneration to the stockholders are very encouraging. The most productive sources of revenue will be furnished by this conveyance; viz. coal, iron, lime, copper, zinc, manganese, copperas, plumbago, turpentine, marble, lumber, manures of various kinds, the products of agriculture, and the fabrics of manufactures." Blurring the line between public good and private benefit, Clinton added: "I should regret exceedingly if this important work should be lost to the public, for the want of three or four hundred thousand dollars. It is manifestly [in] the interest of the stockholders to complete it, and co-operators may confidently calculate upon certain and ample returns for their advances. The estimate of the engineer has been verified by the prosperous progress of the works, and there is not a shadow of doubt as to the resulting advantages to individuals, and as to the immense benefits to the community."[30]

Clinton's advocacy for specific canal-building projects continued posthumously. Just a month before he died, he dispatched a letter extolling the virtues of the Hampshire and Hampden Canal Company's plans to build a canal that would link New Haven (at the mouth of the Quinnipiac River) with the upstream stretches of the Connecticut River all the way to Barnet, Vermont. The editor of the *New-Haven Herald* published this letter, eulogizing Clinton as a technically

astute political leader: "The opinion and estimates of this scientific and disinter-
ested person, whose experience in canalling operations was not inferior to that of
any man now living, and who could not have been biassed [sic] by any interested
motive whatever, are worthy of the greatest deference and respect."[31] In a similar vein, another testimonial (published as "the very last message" Clinton ever
wrote) gave assurances for the commercial prospects of yet another New Jersey
canal venture (the Delaware and Raritan Canal): "As the last emanation of that
departed and immortal mind, on a subject in which it delighted to employ its
energy for the blessings of posterity and the aggrandizement of his native state,
it should be regarded with profound interest, and felt with all the weight of an
oracle by the legislature of New-Jersey and the citizens of the state."[32]

     In sum, it is imprecise to say that the United States in the 1820s was swept by
canal fever, as if some invisible agent carried that infection. Rather, extensive
patches of countryside were torn asunder by thousands of miles of trenches because people believed the promises of wealth they were fed. The credibility of
such promises was immeasurably enhanced by those public and private promoters who were able to persuade DeWitt Clinton to lend his infectious enthusiasm
and learned reputation to the support of canal projects in their states. Not all
canal projects actually turned out to be as profitable as Clinton's many admirers
and imitators were prepared to believe. Expertise was certainly easier to come
by after the Erie Canal's success, and construction costs could even be reduced
in some cases, but basic facts of topography and water supply are not susceptible
to rhetorical persuasion. Most of the canals begun in the 1830s and 1840s barely
survived long enough to reach completion. Some were practical failures because
they were never designed to handle the demand for traffic capacity or rapidly
changing expectations about speed of travel. Others were doomed because they
involved outrageous attempts to scale mountain ranges in the absence of adequate water supplies. In the end, the entire American canal system succumbed
to direct competition with an even more heavily subsidized infrastructure. Railroads proved, in the long run, to be considerably less vulnerable to the dictates of
physical geography.[33]

     Though the heyday of canals was relatively short-lived, their influence on the
shape of the country and on the patterns of its settlement should not be under-
estimated. Canals made New York the number one port of entry to the United
States. Sleepy villages like Rochester, Buffalo, Cleveland, and Akron sprouted
into major cities because of their proximity to the flood of human and commercial traffic channeled by the creation of the Erie Canal and the Ohio and Erie
Canal. Other naturally well-situated towns, like Canandaigua, New York, and

New Philadelphia, Ohio, would remain small forever because these particular canal routes passed them by.[34] Indeed, this pattern of historical contingency helps to illuminate why the mosquito-ridden swamp located at the head of Lake Michigan should ever have grown into the continental interior's greatest metropolis. That Chicago was established at all followed only from the Illinois canal commission's decision in 1830 to hire surveyor and engineer James Thompson to begin to lay out a potentially expansive town grid in that precise location "so that lots could be sold to finance the Illinois canal from Lake Erie to the Mississippi River."[35]

Finally, the process of professionalization of the natural sciences and engineering in the United States was yet one more consequence of the spread of canal fever from New York State to the rest of the country in the 1820s. Fascinating scientific discoveries comparable to those celebrated by DeWitt Clinton and his engineers soon cropped up in virtually every other canal project. The density of these discoveries and their influence on natural history knowledge and engineering practice cannot be overstated. The *American Journal of Science* covered the entire spectrum of science and technology, ranging from reports about the therapeutic value of rhubarb, to essays on theoretical chemistry, mechanics, and physics, to a description of a steam pump. Astoundingly, out of the twenty-five topical scientific articles published in the July 1828 issue, four were devoted to scientific observations of North American natural history made at canal sites: "On the Fossil Tooth of an Elephant, from Lake Erie, and on the skeleton of a Mastodon, from the Delaware and Hudson Canal," by Jer[emias] Van Rensselaer, M.D.; "Notice of the Louisville and Shippingport Canal, and of the Geology of the Vicinity," by Increase A. Lapham, Assistant Engineer; "Geological Nomenclature, Classes of Rocks, &c.," by Prof. Amos Eaton; and "Account of the Welland Canal, Upper Canada," by William Hamilton Merritt, Esq, Superintendent.[36]

Jeremias Van Rensselaer, another cousin of the Patroon, was an accomplished New York naturalist in his own right. His early interest in the location of salt and gypsum led him to develop in 1823 a theory of seashore retreat to explain the deposition patterns of these valuable minerals.[37] His *Lectures on Geology* (1825) is now regarded as "one of the first textbooks on geology published anywhere."[38] In 1826, he reported on the mastodon whose remains were found in the excavation of Cadwallader Colden's New Jersey canal bed.[39] Keen curiosity about natural history might have run in the family, but economic interests could also work hand in glove with political and cultural interests, especially when it came to harvesting wealth from educated study of the past history of the land. New York State's singularly successful demonstration suggested that this was, after all, a fruitful activity.

Among the four articles listed above, Eaton's notice on geological nomenclature was particularly important. Claiming the power to assign names to natural
objects was a critical aspect of science's contribution to the development of an
American national identity. As the founding professor of the Rensselaer School,
Eaton was asserting more than a general tenet of science when he wrote, "Geology is chaos without systematic arrangement." He was seeking to build a case for
a distinctively North American systematic geological nomenclature. Well aware
of European methods, Eaton was happy to jettison any British and French rock
names that reeked of local idiosyncrasy. In fact, he was steeling himself to challenge widely shared assumptions about the general applicability of European
stratigraphy to places outside of Europe. At issue was the question of the universality of sedimentary rock sequences. At stake was the authority of the field investigator versus the authority of the remote system-building theorist in deciding what
names to assign to rocks found in previously unexamined geological localities.

Eaton embodied contemporary paradoxes. Earth science was necessarily becoming a cooperative international enterprise, but all forms of cultural achievement in the United States were heavily invested with nationalistic significance during the decade when the youthful republic had celebrated just its fiftieth anniversary
of independence. These historical circumstances influenced the debate over scientific practice regarding localized geological nomenclature, and it is useful to reex-
amine Eaton's aspirations on behalf of systematic geology through the lens of his
personal experiences and patriotic feelings. Both his unpublished journals and
his published reports give shape and flavor to his profoundly independent-minded
attempts to develop a comprehensive system of North American rock names and
sequences. Eaton's conviction to develop his own nomenclature was born during
his second geological tour of the Erie Canal route. Just two and a half weeks into
the tour, Eaton stopped in Utica and took stock of all his accumulated thoughts
regarding the "basis rock types" to be found underlying the jumble of marl and
alluvion across New York State.

Eaton's journal entry for 30 May 1823 reads: "On reviewing my notes and
remarks upon rocks, I find much difficulty in naming them according to the
nomenclature of any European system. How will it do to adopt the following for
the transition and secondary?" He began, like most other American geological
fieldworkers of the early nineteenth century, by identifying the sedimentary strata
he encountered according to Wernerian principles (see the rock names in Figure
4). As he had done before with his *Manual of Botany*, which was a user-friendly
field guide to the native flora of New York and New England, Eaton intended to
produce a practical guide to the bedrock geology of the same region. The rock

TABLE 1

Amos Eaton's Rock Nomenclature Table of May 1823

Rock Classification from Scratch
Transition
– Argillite
– Calciferous sandstone
– Metalliferous limestone
– Shell limestone
– Graywacke, with its subordinate rubblestone?
– Old red sandstone, with its subordinate conglomerate
Secondary
– Millstone grit, with its subordinate conglomerate shale and coal
– Calciferous sandstone, with its subordinate brittle shale
– Ferriferous slate, with its subordinate argillaceous spar, jasper, iron ore
– Ferriferous sandstone
– Calciferous slate, with its subordinate gypsum, water limestone, and newest shell limestone
– (perhaps swinestone)

Source: Eaton, journal entry dated 23 May 1823 [Utica], *Geological Journal B*, 33–34, box 2, Amos Eaton Papers, New York State Library, Albany. The rock layers proceed from lowest (oldest) to highest (youngest).

classification categories he initially brought to the field were therefore generic (not focused on peculiarities of locale), and at this point his identifications were still primarily defined by observable mineralogical characters, as opposed to relying heavily on the fossil content of any encased "petrifactions." In his stratigraphic practices, Eaton consciously followed English geological models, and he openly professed his desire to build upon established European knowledge and to extend its range of application. In his 1824 report to Van Rensselaer, Eaton laid forth the assumptions that governed his initial attempts at a systematic nomenclature: "I have adopted the European name in every case where an *established name* would apply; but where I have found no settled name which is applicable to the rock under consideration, I have given it a *descriptive* name, in imitation of Bakewell's *metalliferous* and Conybeare and Phillips' *saliferous* rocks." He went on: "To identify the secondary rocks of our district with those of Europe was, from the beginning of my examinations, a principal object."[40] But after having completed so much fieldwork, Eaton had developed serious doubts that this derivative approach could accurately describe all North American rocks: "Several European writers have complained of their contracted strata, so made by the approximation of their primitive ranges; whereas our secondary rocks, along the canal line, are several hundred miles in extent, and remarkably uniform in their leading characteristics."[41]

TABLE 2

Amos Eaton's Revised Rock Nomenclature Table of August 1823

Secondary Class of Rocks

1. Millstone grit
   Subordinate, Conglomerate   Westmoreland, near Utica
2. Calciferous sandstone
   Red sandstone
   Grey & [?] sandstone   Vernon to Genesee River, to near Lock Port,
   Red slate   Ontario Lake &c.
3. Ferriferous slate
   Subordinate, Argillaceous iron ore   Verona, Vernon, Genesee Lower Fall
4. Ferriferous sandstone
   Subordinate, Argillaceous iron ore   Vernon, Verona
5. Calciferous slate
   Subordinate Gypsum beds
   Shell limestone   Manlius, [illegible], Genesee River
   Water limestone
6. Geodiferous limestone   Rochester, Lock Port, Niagara Falls
7. Hornstone lime rock   Cayuga Lake below Ithaca, Black Rock
8. Pyritiferous slate
   Subordinate, Pyritiferous limestone   Ithaca, 18-Mile Creek in Lake Erie

Source: Eaton, journal entry dated 14 August 1823 [Niagara Falls], *Geological Journal C*, 90–91, box 2, Amos Eaton Papers, New York State Library, Albany.

Eaton's attempts to comprehend the system of New York's rock layers were shaped by his persistence, by the fact that he was working under pressure, and by his hunger to be acknowledged as the preeminent geologist of the northern United States. The pressure was somewhat self-imposed, since he had contracted to perform a geological survey report of the Erie Canal route for Van Rensselaer the previous summer.[42] Shortly after returning to his home in Troy from the canal tour in May 1823, Eaton misplaced the book containing all his notes and observations, including the original chart of rock names (reproduced in Table 1). Determined to reconstruct his notes, Eaton returned to the canal route as soon as he was free of summer teaching duties at Troy's Female Academy. For five and a half weeks in July and August 1823, he reexamined the 350-mile tract between Troy and Lake Erie and tried once again to make sense of the confusing formations. At Niagara Falls, he stopped to review all the secondary rocks he had seen and to catalogue the characteristic localities where he had found them. The freshly begun volume of his geological journal displays the list reproduced in Table 2. The term *geodiferous* exemplifies the license Eaton took to coin new geological terms. West of the Genesee River, he discovered that geodes abound in the stratum that occupies the space between underlying slates and overlying limestones.

After enduring five weeks of remote field geology, travel, and midsummer heat, Eaton's health suddenly failed. Then in his late forties, Eaton collapsed upon his return to Troy on 28 August. While recuperating during the first days of September, Eaton forged ahead with his work, piecing together a rudimentary geological "section" of the canal route. From mid-September to mid-November Eaton continued to suffer poor health, complaining afterward of having spent four of the weeks laid up by "inflamed lungs" and "lake fever." On 17 November, upon his return from teaching a course to medical students in Vermont, Anna Bradley Eaton (his third wife) greeted her husband with unexpected good news: she had found the lost journal containing the May 1823 tour notes.[43] Eaton prepared immediately to make one more tour of the canal route. Confident that he now knew precisely what gaps remained in his understanding and where he might seek answers to these questions, Eaton left the very next day. He covered two hundred miles in just one week, observing rock localities during the daytime and traveling by packet boat at night, until ice prevented his progress further west on 24 November 1823. As we saw in chapter 5, these were the trips that yielded his initial canal report and geological profile, both published early in 1824.

Eaton revised and republished his classification system in 1829, after which the nomenclature issue ultimately caused a bitter personal conflict between Eaton and another scientist. Surprisingly, it can be argued that Eaton's greatest achievements (the systematic nomenclature and the first geological map of New York State), as well as his most deliberately provocative assertion (that more of the Earth's history lay visible and well displayed within the territory of New York State than in the haphazardly scattered beds and formations of all of Great Britain), all followed directly from a petty argument with a transplanted Englishman named George William Featherstonhaugh, who was then in the midst of enjoying a relatively long, successful career as a professional geologist in the United States. After brief visits to New England, Featherstonhaugh had made his residence in the United States in 1809, just as hostilities between the republic and its mother country were heating up (leading to the War of 1812). During the 1820s, he was an important organizer of New York's lyceum and a protégé of Stephen Van Rensselaer.[44] Having built up his reputation as a leading geologist in the United States, he nevertheless retained a lifelong allegiance to the interests of Great Britain, and in 1839, as a British appointee to the boundary commission, he worked to justify the stingiest sustainable American territorial claim.[45] Not long afterward, he left the country, and negotiations of the boundary between Maine and New Brunswick were settled by the Webster-Ashburton Treaty of 1842.

From the beginning, the Englishman had a penchant for baiting Americans into rude displays of their national feeling: "Like many other newly arrived Englishmen he seemed to be unconsciously offending his hosts . . . [by speaking] of British superiority in various things until [on one occasion] this gentleman asked him if it was true that the moon 'was larger and rounder in England.'" Despite such initial faux pas, Featherstonhaugh managed to marry into the socially prominent and wealthy Duane family, whose extensive estates lay immediately to the west of Albany, and he soon established himself among upstate New Yorkers as an agriculturalist and businessman. Featherstonhaugh's advantageous match also gave him a distant familial connection to Stephen Van Rensselaer. He therefore had an important advantage over Eaton in his social proximity to the Patroon. Their mutual interest in developing institutional support for science began with the New York Board of Agriculture, for which Featherstonhaugh served as the first corresponding secretary. In fact, it was under his auspices that the board had published Eaton and Beck's geological survey of Albany County in 1820. In this first professional contact, Featherstonhaugh praised Eaton's work, but trouble would come once the native New Yorker's scientific reputation began to approach that of the Englishman.[46]

Eaton grew to detest Featherstonhaugh both personally and professionally. The two men rapidly rose side-by-side among the ranks of Albany's scientific-minded citizenry during the 1820s. Both men found a powerful patron in Van Rensselaer, and both mobilized his sponsorship to elevate themselves to national recognition for geological research by 1830. When Van Rensselaer entered into congressional politics, he considered Featherstonhaugh the "Hercules in Agriculture" and heartily recommended his sheep-farming colleague to the most influential men in Washington. Equipped with this generous introduction, Featherstonhaugh ingratiated himself with prominent southern gentlemen such as James Madison, James Monroe, and Henry Clay. The ambitious Englishman hinted broadly that he was far better equipped to serve as a federal expedition geologist than any American could possibly be.[47]

It is not entirely clear if and when Featherstonhaugh suffered his own break with the Patroon. Featherstonhaugh's enthusiastic support for Clay in the 1824 presidential election meant that he would have had no problem with Van Rensselaer's deciding vote for Adams, which followed Clay's deal to assist Adams in return for the office of secretary of state. Perhaps the Englishman simply felt that he had ascended to a position of national importance far beyond the point where the Patroon's parochial New York power mattered. From Eaton's perspective, there was no doubt that Featherstonhaugh had totally squandered the Patroon's favor by 1829, when, Eaton alleged, Van Rensselaer asked, "Eaton do you know of any weed that will serve as an antidote for a Featherstonhoughean incubus?"[48]

The validity of this testimony is, of course, complicated by Eaton's rivalry with Featherstonhaugh, which at that moment was reaching a fever pitch.

Eaton's views on the rules for innovating in scientific nomenclature would harden into a position grounded as much in the arousal of his patriotic ire as in his extensive field experience. He had published several draft renditions of his new nomenclature during the previous year, and in the summer of 1829 he was testing the waters for the full-fledged diluvial system as he prepared his *Geological Prodromus* for publication. Featherstonhaugh decided to launch a preemptive strike, ridiculing Eaton's new system of New York's strata as being locally derived, foolish, and wrong. William Buckland's biographer Nicolaas Rupke suggests that Featherstonhaugh may have had a nationalistic motive. By the late 1820s, Eaton had incorporated fossil-based stratigraphy more prominently into his thinking; according to Rupke, this telltale evidence of French stratigraphic influence lay at the heart of Featherstonhaugh's attempts to contemptuously discard the American's proposals.[49] In any case, two important points need to be made about the Eaton/Featherstonhaugh brouhaha. First, contrary to a view that became popular among historians of American science in the late twentieth century, Featherstonhaugh's attack on Eaton was an exception—not representative of any widespread contemporary critical attitude towards Eaton's theoretical attempts at system building.[50] Second, it is a mistake to suppose that Featherstonhaugh singled Eaton out for this kind of rebuke. The Englishman had a long track record of challenging the credibility of North American geologists whenever they offered their own names for rocks, insisting that they did so because they were simply ignorant of the strata of Europe. When Featherstonhaugh finally published his own views on the equivalence of strata between the continents in 1832, he predictably took the British nomenclature and specific order of superposition deemed by William Conybeare to be the standards by which North American rocks ought properly to be named and located.[51]

Unlike some of his countrymen, who could not help but be impressed and chastened by Featherstonhaugh's airs of superiority and claims of personal intimacy with Europe's leading geologists, Eaton called the Englishman's authority hollow and had the audacity to advocate for his own. Before angrily exploding in public, however, Eaton tried to learn what had provoked this outrageous put-down in a private conversation with William Cooper, a New York attorney who was on good terms with Featherstonhaugh. Eaton explained his outrage to Cooper: "As the single principle . . . upon which the whole superstructure of the science of Geology is erected is, — 'that the *order* of superposition in the arrangement of strata is universal;' to say that any one, making any pretensions, denies this principle, is to say, that he is ignorant of the very first rudiments of geology. The publication in question, which I consider is holding me up to the scientific world in a ridiculous point of view, certainly assigns to me this full measure of ignorance."[52] By reframing the challenge of establishing international equivalents among the different systems and the various sequences of strata that had been documented in different parts of the world, Eaton was among the first Americans to boldly assert a fundamental sense of independence. Just as the Revolutionary War had been fought so that American social mores and political structures should be free of compulsions to replicate the corruptness of Old World monarchies, Eaton now insisted that American geology should suffer no obligation to conform slavishly either to European rock patterns or to European-invented scientific models. Ironically, the predominant thrust of Eaton's forthcoming *Prodromus* was his effort to highlight those analogies and consistencies that he had discovered between the strata of both continents. When Eaton met Featherstonhaugh later that same day in the company of John Torrey, William Cooper, and some other members of the New York Lyceum, he attempted to smooth out the misunderstanding by stressing that congenial aspect of his work. But Featherstonhaugh was in no mood for compromise or collegiality, and Eaton came away feeling even more painfully stung. He wrote sarcastically in his private journal:

> It is surely ridiculous to give local names to rocks, and strata of detritus, supposed to be general. I beg that American geologists never give in to this absurdity. Let us apply this rule to see its absurdity. In Columbia county N.Y, there is a rock, formerly much used for fire stones. It is the 2d graywacke, becoming sandy. But it is found [in England] only in a place formerly called Dodgingtown, now Pilfershire. —this must be called Dodgingtown stone *being the oldest name*. A vermicular variety of lias was first observed in a place called Fuddletown—this must be Fuddletown rock &c. Thus we must have a shamefully ridiculous nomenclature like this: Bagshot sand, Purbeck limestone, Fuddletown rock, Kimmeridge clay, Dodgingtown stone &c. Europeans are too jealous of each other to adopt a name not taken from [their own] quarry men; even where there are two names for the same rock. Have we become so ridiculous?"[53]

Having finally expended the greater part of his vitriol, Eaton reiterated his rationale for American self-reliance in science: "I am willing to follow Europeans in every thing laudable, and am an enthusiastic admirer of their rapid progress in geological discoveries. But, while we have independence enough to reject their political absurdities, shall our men of science suffer their aspiring minds to be shackled by the low envy of foreign speculators of science[?]." But the sour taste of Featherstonhaugh's accusations could not be so easily washed away: "A swag-gering European braggadocio comes here to dictate to our patient enquiries after scientific truths. His absurdities must be swallowed (digested if possible) and truth must flee before him. Shall we endure this? Or shall we separate the wheat from the chaff, and erect a standard of our own in Science as well as in government? Let us profit by foreign improvements; but let us resist a ridiculous nomencla-ture, founded on envy in some cases, and local pride in others. Above all let us distinguish between *European* visitors who are modest enquirers after truth, and the swaggering ignoramus, who talks largely and knows nothing." Eaton closed his diatribe with a call to both American nationalism and self-reliance in geology: "Let us discard the ridiculous nonsense of those who wish to make us believe, that we must see Europe, or cannot be geologists. Why not say with DeLuc, that to be geologists, Europeans must see America, where Nature has performed his work on an enlarged scale. Is Buckland a geologist? Is Cuvier, Brongniard [*sic*], or Bakewell a geologist? Is McCulloch, Jameson, or Conybeare, a geologist? Neither of these gentlemen ever saw America, which is the only country in the world where general strata are of sufficient extent for determining their order of superposition or for distinguishing between extensive beds and general strata."[54] Eaton was clearly carried away by his temper, for he usually professed high regard for the works of both Buckland and Cuvier.

By 1830, Eaton would justify and promote his rock-naming practices: "Who-ever is 'first in the field' of natural science, has an exclusive right *to give names*. His successors should either adopt his names, or give them as synonyms and equivalents. This is essential to the very being of science. But English and French geologists have introduced new names, not adopted in Germany; because new discoveries have made them necessary. I have done the same thing in America, and for the same reasons."[55] He presses the point home even more succinctly in another passage from the first edition of his *Geological Text-Book*: "The little island of Britain can furnish no authority for a *general system*."[56] Eaton would eventually accept the equivalence of prevailing English systems and teach the European equivalents to his Rensselaer classes. Still, he remained proud of hav-ing developed an independent North American nomenclature based on the rocks of New York State. The prefatory notes to his *Geological Note Book, for the Troy Class of* 1841 clearly articulate his lack of repentance:

> My old students will be surprised at my adopting the Cambrian, Silurian, and Devonian systems of Sedgwick and Murchison. But they must be reminded, that the *Cambrian* is my first graywacke, (as published in 1824) with most of

its associates—the *Silurian* is my second graywacke, after giving cover to the Cambrian, the metalliferous lime-rock—the Devonian is my subordinate red sandstone group. When I published this distribution, it was rejected by English geologists, and called an unnecessary novelty by Buckland. Now my 'unnec-essary novelties,' published seventeen years since, in my report to the Hon. Stephen Van Rensselaer, seem to have been recently applied to the rocks of Wales and Devonshire, by two of the most active geologists in England. I was right, because the *rocks* were there; but I was deficient in that knowledge of foreign rocks, taken in full extent, which was necessary for determining their equivalencies. I then proposed to give descriptive names only; leaving to others the authority of fixing permanent names.[57]

The outcome of their struggle for supremacy, though colorful in its recrimina-tions and ambiguous in its effect on Eaton's own reputation, proved ultimately fruitful for the practice of natural history in New York.

LEGACIES OF THE NEW YORK SYSTEM

Eaton's geological successors were grateful for what he accomplished by setting North American geological nomenclature on its own footing, independent of European dictates. Eaton's most famous student, James Hall, who would serve for five decades as the New York State paleontologist, wrote respectfully of his mentor's first-order attempts to organize tables of strata from scratch and in the ab-sence of very soon-to-be outmoded theoretical and methodological frameworks: "Nearly all the rocks of western New-York are enumerated in the order of suc-cession; and, with some exceptions and omissions, the order is correct, and the subdivisions will always hold good in the science. It is a remarkable fact, that at this early period, Mr. Eaton should have recognized the sandstone of the Catskill mountains as the Old Red [Sandstone] of Europe; which, now that we have identified its characteristic fossils, is proved to be true." Hall speculated further, in a manner that echoes his own later difficulties disentangling the confusingly jumbled layers of the Taconic Mountain strata: "Had he seized this grand idea, and confined himself to the elucidating of the strata below the Catskills, he would have brought to light the most interesting series of rocks yet known in any part of the globe." But Eaton's great global ambition was also the signal reason for his weakness, according to his student: "The great source of error throughout seems to have been the prevailing desire to identify, within the limits of New-York, all the rocks and systems published in Europe from the Tertiary downwards."[58]

John Mason Clarke, Hall's successor as state paleontologist, commented that

Eaton's 1832 edition of the *Geological Text-Book* represented a premature attempt to integrate as well as supplant European classifications for rock formations. The revolutionary theoretical implications, as understood from Clarke's perspective a full century later, may have been well beyond Eaton's imagination, but the impulse to champion locally derived rock names was at least a step toward understanding the dynamics of regional sedimentary rock formation: "[Eaton's classification scheme for rocks and detritus] contained the germ of an important conception; that of cycles of rise and fall of the sea bottom indicated by repetitive succession of coarse to fine and fine to coarse. Indeed this seems to have been the first enunciation of a principle which is now fundamental to accepted interpretations of stratigraphical succession, and as we look back over the historical development of the science it is to justify Eaton in the recognition of geological succession so controlled so as not to fit the European categories." In a way, Eaton suffered the ignominious fate that must befall all scientific innovators who break promising new ground.

As compared to the improved systems that would follow, Eaton's original attempts appear naive and simplistic. In Clarke's estimation, history's ultimately positive verdict would require time because "[Eaton] failed to fortify his propositions, his contemporaries reprobated them and his successors ignored them; until John S. Newberry, a half-century after, brought forward a clearer definition, and Charles Schuchert interpreted it in terms of changing continental shelves."[59] In 1899, Clarke and Schuchert collaborated to fulfill Eaton's vision by compiling and publishing their own rendition of the nomenclature of New York's geological formations in *Science* (still the leading journal of American scientific discourse, though many discipline-specific rivals had been launched by that time). Their purpose was unabashedly to standardize and promote "a set of designations derived from characteristic localities of the New York paleozoic, and thus to preserve, under the necessity of change, the eminent title of New York State to its full and ancient representation in the classification of the paleozoic deposits and time."[60] The New York system, as set forth in their authoritative table, retained such Eaton-flavored equivalences in era names as "Cambric or Taconic," "Champlainic (Lower Silurian and Ordovician)," and "Ontario (or Siluric)." Even within the Devonic Era (thus accepting the universal equivalence with southwestern English rocks first studied by Eaton's contemporaries Roderick Murchison and Adam Sedgwick), Mason and Schuchert listed rock groups based on New York locales (ranging from Helderbergian and Oriskanian up to Senecan and Chautauquan).

When all was said and done, the debate that Eaton had launched seventy years earlier about localized geological nomenclatures came down to this handful of significant questions:

- Are New York and Great Britain, for example, distinct geological "texts" to be read, or are they parts of the same universal "text"?
- With what language should geologists attempt to read these texts? In other words, how is it possible, and why is it necessary, to develop a universal nomenclature, regardless of one's answer to question 1?
- If one locality is to be privileged over another in the generation of a piece of the universal nomenclature, does it matter which locality gives birth to the characteristic description of a formation?

Reviewing Amos Eaton's life and career as an originator and promoter of geological practices provides some intriguing answers to these questions, which turned out to have consequences not only for the character and practice of geology as a global science but also for the advancement of the aspirations of other "hinterland" scientists who sought to elevate appreciation for their own countries as places of importance. Careful study of Figure 4 shows how Eaton integrated his fundamentally Wernerian approach to mineralogy with the discoveries he had made of sunken forests, boulders, and commercially valuable materials along and beneath the Erie Canal.[61] While he did not champion local names for rocks, the rationale he advanced regarding American subservience to foreigners in science implies that he was fully aware of the nationalistic dimension of his theoretical work. Buckland, Cuvier, and Bakewell had never seen these rocks. By challenging even his heroes, Eaton effectively framed a set of standards of research practice that promoted American geology.

Granted a second chance in life, Eaton managed to launch a successful scientific career. After gaining early notoriety as a popular botanist, he became an influential, if controversial, geological observer and taxonomist. The assembly of a compact, unified, independently derived system of geological strata was in some ways comparable to his achievements in botany, but the nagging rhetorical questions Eaton peppered throughout his geological journals reveal that he was driven far beyond practical matters to grapple with fundamental philosophical and methodological principles. In so doing, he arrived at a number of related organizing principles for the conduct and instruction of the sciences. Specifically, geological understanding should be driven by experience in the field, not the ivory tower. The field where one might obtain the most useful experiences, in turn, should be one where phenomena are grand and comprehensive, not one that is cramped or confused. In Eaton's view, therefore, a field geologist working

CASE OF SPECIMENS, Classes 6 & 6.

No.	GENERAL DEPOSITES AND SUBDIVISIONS.	VARIETIES.	IMBEDDED AND DISSEMINATED SUBSTANCES.
26.	SUPERFICIAL ALLUVIUM. / B. Granulated, (from gray-wacke.) / A. Clay-loam, (from argillite.)		Various boulders. Pebbles.
25.	STRATIFIED ANALLUVIUM. / C. Lias. / B. Ferriferous. / A. Saliferous.		Gypsum. Shell limestone. Reddle.
24.	POST-DILUVION. / B. Sediment. / A. Pebbles, (in the rocky bed of a river.)		Various boulders. Trees and herbs. Fish bones and shells. Works of art.
23.	ULTIMATE DILUVION, (on crag in oolit forests.)	Yellowish grey. Greyish yellow.	
22.	DILUVION, (in an antidiluvial trough.)	Quicksand. Gravel. Vegetable mould.	Boulders. Trees and leaves. Bones and shells. No works of art.
21.	ANTIDILUVION. / C. Bagshot sand and crag. / B. Marley clay. / A. Plastic clay.	Quicksand. Yellow sand. Hardpan. Brick earth.	Pudding-stone. Bolbstone. Bure. Shell-marle. Indurated marle. Septaria?

CASE OF SPECIMENS, Classes 4 & 3.

No.	GENERAL STRATA AND SUBDIVISIONS.	VARIETIES.	IMBEDDED and DISSEMINATED.
20.	BASALT. / B. Greenstone trap. (columnar.) / A. Amygdaloid, (cellular.)	Granular. Compact. Trandstone.	Amethyst. Calcedony. Prehnite. Zeolite. Opal.
19.	THIRD GRAY-WACKE. / B. Pyritiferous grit. / A. Pyritiferous slate.	Conglomerate, (breccia.) Calcareous grit. Old red sandstone. Red-wacke. Argillaceous.	Grindstone. Hornstone? Hornslate. Bituminous shale and coal. Fibrous barytes.
18.	CONGLIFEROUS LIMESTONE. / B. Shelly. / A. Compact.		Hornstone.
17.	GYPSIFEROUS LIMEROCK. / B. Sandy. / A. Stariactitar.	Foetid.	Snow-Gypsum. Strontian. Zinc. Fluor spar.
16.	LIAS. / B. Caleiferous grit. / A. Caleiferous Slate.	Shell grit. Argillaceous. Concuthoidal.	Shell limestone. Vermicular. Water cement. Gypsum.
15.	FERRUGINOUS ROCK. / B. Sandy. / A. Slaty.	Conglomerate. Green. Blue.	Argillaceous iron ore, (reddle.)
14.	SALIFEROUS ROCK. / B. Sandy. / A. Marle-slate.	Conglomerate. Grey-sand. Red-sandy. Grey slate. Red slate.	Salt, or salt springs.
13.	MILLSTONE GRIT. / B. Conglomerate. / A. Sandy.		Coal.

CASE OF SPECIMENS, Classes 2 & 1.

No.	GENERAL STRATA AND SUBDIVISIONS.	VARIETIES.	IMBEDDED and DISSEMINATED.
13.	SECOND GRAY-WACKE. / B. Rubble. / A. Compact.	Red sandy, (old red sand?) Stone-slate. Grind-stone.	Manganese. Anthracite.
12.	METALLIFEROUS LIMEROCK. / B. Shelly. / A. Compact.		Birdseye marble.
11.	CALCIFEROUS SANDROCK. / B. Greedifrous. / A. Compact.	Quartzose. Pisary. Oolitic.	Semi-opal. Anthracite. Barytes. Concretitic concretions.
10.	SPARRY LIMEROCK. / B. Slaty. / A. Compact.	Checkered rock.	Chlorite. Calc spar.
9.	FIRST GRAY-WACKE. / B. Rubble. / A. Compact.	Checkered rock.	Milky quartz. Calc spar. Anthracite.
8.	ARGILLITE. / B. Wacke Slate. / A. Clay Slate.	Chloritic. Glazed. Roof-slate. Red. Purple.	Flinty slate. Anthracite. Striated quartz. Milky quartz. Chlorite.
7.	GRANULAR LIMEROCK. / B. Sandy. / A. Compact.	Verd-antique. Dolomite. Statuary marble.	Tremolite. Serpentine. Chromate of iron.
6.	GRANULAR QUARTZ. / B. Sandy. / A. Compact.	Ferruginous. Yellowish. Translucent.	Manganese. Hematite.
5.	TALCOSE SLATE. / B. Fissile. / A. Compact.	Chloritic.	Octahedral crystals of iron ore. Chlorite.
4.	HORNBLENDE ROCK. / B. Slaty. / A. Granitic.	Greenstone. Gneissoid. Porphyritic. Sienitic.	Granite. Actynolite. Augite.
3.	MICA-SLATE. / B. Fissile. / A. Compact.		Staurotide. Sappare. Garnet.
2.	GRANITE. / B. Slaty, (gneiss.) / A. Chrysolitic.	Sandy. Porphyritic. Graphic.	Shorl. Plumbago. Sienite. Diallage.

in New York State was automatically better qualified to make pronouncements about what generally lies beneath the Earth's surface than was any European arm-chair natural philosopher. Eaton had no patience for arrogant interlopers; foreign visitors at best could become auditors of American geological expertise. Eaton may have been recalling the notoriously presumptuous Constantine Rafinesque, who somehow managed to exercise uncharacteristic restraint when he accompanied Eaton's Rensselaer Institute field trip in 1826. This episode earned the foreigner a cordial, collegial reception and repaired an old wound; Rafinesque had dismissively criticized the first (1818) edition of the Eaton's *An Index to the Geology of the Northern States*.

By contrast, the denouement of Eaton's showdown with Featherstonhaugh was not at all cordial, but their competition for national and international prestige nevertheless yielded important results whose implications reached far beyond their personal reputations. Their altercation erupted precisely at the moment when New York State began to address the question of how to mobilize public support to carry forward the scientific work that Van Rensselaer had privately underwritten. In his final annual message to the legislature on 1 January 1828, Governor Clinton signaled his interest in helping New York to develop its native supply of coal. In the words of historian Michele Aldrich, "Clinton knew that geologists had suspected coal should be found within New York, given its stratigraphic column's similarity to the European rock sequence." Eaton was, of course, primarily responsible for misleading his contemporaries by relying too heavily on Wernerian assumptions about the universal superposition of secondary strata. Aldrich implicitly acknowledges this error, while deflecting the blame to include others: "The evidences [the legislative committee] cited for the presence of coal were the stratigraphic identity of the sandstones along the Erie Canal with the rocks overlying the coal measures of Europe—a mistake based on the use of mineralogical similarities rather than fossil identification—and the chemical analyses of Tioga county coal made by William Meade."[62] Had Eaton's early interest in fossils been more highly developed, he too might have recognized the bitter truth: Eaton's saliferous rocks were not indicative of a third cycle of "graywacke" deposition.[63] In other words, New York had no coal of its own. The layer containing Pennsylva-

Figure 4 (opposite) Amos Eaton, "Tabular View of North American Rocks," 1829. These three tables originally appeared as the three frontispiece pages preceding the text proper of Eaton's *Geological Prodromus* (Albany, 1829). His descriptive nomenclature generally derived from Abraham Werner's system of geology based on mineralogical features, and his list of formations was numbered from oldest to youngest.

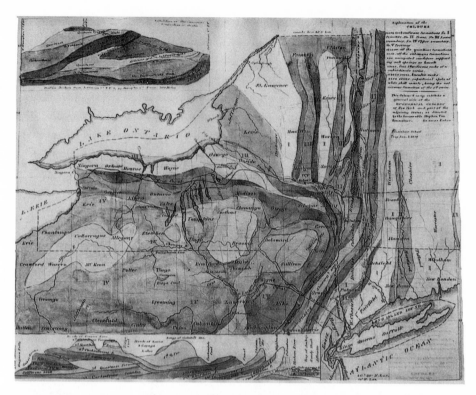

Figure 5 Amos Eaton, "Map of New York Economic Geology," 1830. This map constitutes the first detailed attempt to combine "every thing of importance . . . relating to the mineralogical resources of this district [New York State]" together with systematic representations of stratigraphy (profiles) and surface bedrock territories across the entire state. Across the map's bottom, Eaton showed subsurface rock relationships from Niagara Falls in the west to Long Island in the east. Along the upper left he similarly depicted how many of the same rock layers were stacked from the Adirondacks in the north to the Pennsylvania border in the south. By including all these different representations together on the same map, Eaton showed how New York geology embodied his comprehensive theory of Earth's history. Quotation from Amos Eaton, "Geological Prodromus," *American Journal of Science* 17 (1830): 68–69. *a.* Thanks to Craig Williams, curator of photographic and graphic materials at the New York State Library, for giving me my first slide reproductions of this map; and to Gerald Friedman, who allowed me to take a snapshot of an original print from his collection of maps at the Northeastern Science Foundation in Troy. This high resolution photograph of Friedman's print was kindly provided by Michele L. Aldrich and Alan Leviton. *b.* Cartography provided by Bill Nelson.

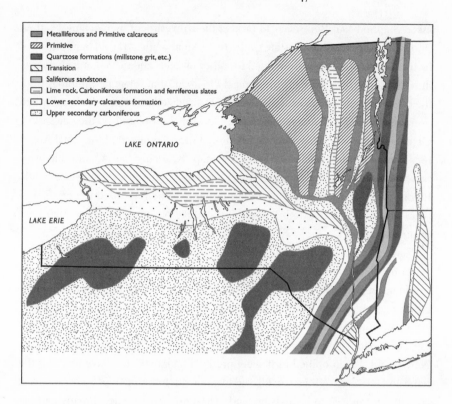

Metalliferous and Primitive calcareous
Primitive
Quartzose formations (millstone grit, etc.)
Transition
Saliferous sandstone
Lime rock, Carboniferous formation and ferriferous slates
Lower secondary calcareous formation
Upper secondary carboniferous

LAKE ONTARIO

LAKE ERIE

nia's beds of coal would never be found below the ground but instead correlated to layers of rock far higher in the order of superposition. If they ever existed north of the state boundary, they were now totally eroded away. Optimistic protestations to the contrary did not give Eaton much insurance against eventual public disappointment and recrimination, especially as they were so instrumental to his call for a state-funded comprehensive natural history survey.

More immediately, however, Eaton saw Featherstonhaugh's application in 1828 for public funds to produce a geological map of the state as an attempt to steal his place in the spotlight by promising a more conservative practical result. Eaton certainly took offense, reacting "with hysterical charges of plagiarism." Despite all of the Englishman's disrespect for Eaton's nomenclature, the fulfillment of his mapmaking proposal would have required extensive reliance on and use of the data Eaton had gathered.[64] Eaton felt strongly that geological sections and profiles (diagrams showing cuts into the Earth's strata along a line) were far more accurate and informative than crudely colored maps. That is why Eaton had never previously composed geological maps for his reports. Now, however, "enraged by what he regarded as an attempt to cash in on his fieldwork, Eaton

asked Stephen Van Rensselaer to protect their investment by blocking Featherstonhaugh's appointment as state geologist."[65] Meanwhile, to avoid being scooped
by his competitor, Eaton suspended his other projects and moved quickly to produce a map of New York's "Economical Geology" that integrated findings from
the extensive fieldwork Van Rensselaer had subsidized and Eaton had performed
over the course of the preceding decade. Devoting himself fully to the new task in
October 1829, Eaton sacrificed completion of the second part of the canal survey
report in order to work fervently on the map. Realizing that a comprehensive
survey was really needed to underpin a map based more on facts than guesswork,
Eaton also sent Van Rensselaer a draft proposal to be forwarded under his auspices to the legislature, urging that the state take on financial responsibility for
future survey work.

Neither Eaton's confidence in the possibility of finding coal somewhere beneath the Erie Canal nor his sense of injury at the hands of Featherstonhaugh
would appreciably diminish by the following year. Indeed, his anger at the Englishman festered so badly that it found its way into the *Geological Text-Book* he
was compiling for his Rensselaer Institute students. In the context of an introductory description of American efforts to piece together a sense of the geology of the
New World, Eaton opined in flowery prose: "Foreigners* too have held out the
hand and tendered us their friendly aid to guide our tottering steps." This apparent
compliment proved to be a savage barb when one read the accompanying footnote, which revealed Eaton's bitter feelings: "I would except the few foreigners
(self-styled geologists) who have come among us, and have endeavored to excite
feuds, without success. These came to this country totally ignorant of the science,
but having caught a mere smattering from our geologists, they began to 'hail from
Europe' as wonders there. As such they obtained admittance for their scurrility in
some of our journals 'of easy virtue;' though no distinguished European geologist
will acknowledge them, or even heard of them as geologists."[66] Two years later,
in the second edition of his *Geological Text-Book*, Eaton dodged the obvious
chance to revise or eliminate his thinly veiled insult to Featherstonhaugh. Instead, although feigning embarrassment about it by "begging" the reader to erase
the remark, Eaton appended a self-righteous explanation: "I regret its publication
exceedingly. I wrote it under a degree of excitement, produced by the remarks
of a friend; and it was in print before I took time to reflect on the absurdity of
introducing such remarks, *however just*, into a text-book for youth."[67]

Regardless of the circumstances, Eaton's map of New York State would establish yet another valuable baseline for comparable achievements in geological
mapmaking. By 1836, when the New York legislature finally approved the plan

and budget for a comprehensive analysis of the natural history of the state (divided into four large districts), eleven other states had already initiated their own publicly funded geological surveys. Though Eaton was by that time too physically limited to engage in any of the work for New York's natural history survey, four individuals he had trained were among those offered key positions (Ebenezer Emmons, Asa Fitch, James Hall, and Edward Hitchcock). Seven more states joined the geological survey bandwagon within the next three years (including Michigan, which lured Douglas Houghton westward with a geologist post). High-profile opportunities to lead scientific expeditions proliferated across the country. Because Eaton had taught a generation of students in his own peculiar hands-on style, elements of the New York system extended even beyond the direct application and refinement of those geological theories and techniques he first systematized while tramping back and forth across the state.

Little of Eaton's eclectic practice of combining traditional Wernerian mineralogy with Cuvierian stratigraphy would dictate the findings of his most capable students, but this fact in no way diminishes his contribution to the establishment of a professional cadre of American geologists. Eaton opened a way forward by the sheer force of his readiness to gather data, offer imaginative ideas, and risk being wrong. His ingenious attempts to formulate a systematic model of diluvial debris deposition expanded upon Buckland's ideas while drawing international attention to the wonders of New York's physical geography. Most importantly, by offering and defending an original North American system of rock nomenclature, Eaton shattered the stultifying effects of European pretensions to global authority in geology.

PART THREE

Entertaining Deep Time
and the Sublime

CHAPTER SEVEN

Literary Naturalists

The canal connecting the ocean with the lakes outdoes the enter-
prizes [*sic*] of Egypt and France. A solemn regard to the education of their offspring will accompany [Americans] wherever they
go. — The rudiments of beneficial literature and art will be instilled
early into their minds. The ornamental branches in due season and
progression. High science will gradually succeed. Every vegetable
will respond to its name, and tell its excellent or noxious qualities.
Ores and rocks shall rise from the bottom of the mine and descend
from the top of the mountain, and arrange themselves in museums.
                              — *Samuel Latham Mitchill, 1821*

Interdependence among science, literature, and the arts was perfectly reasonable to early nineteenth-century New Yorkers. To rigidly separate or divorce these
branches of creative thought would contradict the Enlightenment tendency to
seek affinity (if not unity) among the forms of human intelligence, an outlook
widely shared among the Founding Fathers and that generation of leaders who
followed them. So it comes as no surprise that an early practitioner of American natural history like Samuel Latham Mitchill should espouse the mutually
reinforcing powers of technology, intellect, creativity, and morality. A sense of
urgency and concern, however, lay beneath such exhortations. Predictions of
an imminent blossoming of American sciences and letters echoed in public lecture halls and littered the pages of learned journals throughout the first several
decades of the nineteenth century, as if repeated invocation could conjure its
fulfillment. This same basic anxiety motivated the rapid proliferation of those
New York literary, philosophical, and historical clubs that had provided such
a rich context for DeWitt Clinton's aspirations to intellectual leadership in the
1810s. New York was far from unique in this regard. Cities across the new nation
were filled with individuals who felt the need to prove themselves in terms of

cultivating native talent and developing social institutions that would help foster impressive cultural achievements, especially literary ones.[1]

Just as the scientific investigation of American nature had promised to be a boon to Clinton's political career, so might the wonders and history of the landscape lie at the heart of the cultural impulse to create something new and noteworthy in the fields of fiction, poetry, and prose. A harbinger of this marriage between American natural history and imaginative literary production can even be seen in the mythic potential described by the first famous Hudson River author, Hector St. John de Crèvecoeur. His *Letters from an American Farmer*, written at his 250-acre Orange County farm in 1782, framed the question "What is an American?" in its third letter/chapter. An interesting hint at the answer is buried in Crèvecoeur's first letter, which explicitly links the quest to learn scientifically about American land to the ultimate maturity and greatness of the new nation: "As it is from the surface of the ground which we till that we have gathered the wealth that we possess, the surface of that ground is therefore the only thing that we have known. It will require the industry of subsequent ages, the energy of future generations, ere mankind here will have the leisure and the abilities to penetrate deep and in the bowels of this continent search for the subterranean riches it no doubt contains."[2]

Those New York writers who would follow Crèvecoeur in building an original American literature might have taken his words with varying degrees of seriousness. Influential men of early American letters frequently found the trope of nature to be sincerely inspiring, even as they exhibited a predilection to treat the character of the natural historian as a universally recognizable (often comic) foil to an unlettered Everyman protagonist. Both of these tendencies attest to the raw material that the science of geology offered to nurture an original literature in early national New York. This chapter examines the relevance of natural history knowledge and practices to the lives, careers, and writings of Amos Eaton's three most famous contemporaries in the world of New York letters: Washington Irving, William Cullen Bryant, and James Fenimore Cooper.

INVENTIVE HISTORIES: WASHINGTON IRVING

In 1802, while clerking under judge Josiah Ogden Hoffman in New York City, Amos Eaton befriended Washington Irving. Seven years younger than Eaton, Irving was a fellow law student whose remarkably fruitful literary imagination would carry him astray from a dreary life in the courtroom, just as Eaton's scientific curiosity eventually would divert him. Pressed with the obligations of a wife and child at that time, however, Eaton gained quick admission to the bar and launched his own practice upstate. The colleague he had left behind in New York wrote to him, confiding negative feelings about the profession they had both chosen: "Oh if I could but have my wishes gratified . . . how far behind me I'd leave this wrangling, driving, unmerciful profession. A little independency, a snug handsome little wife (who I have in my eye and in my heart), and an honest friend or two . . . but here I am building castels [sic] and picturing distant scenes that will never be realized."[3]

Unlike Eaton, whose social advantages stagnated in the years leading up to his imprisonment, Irving was blessed with many opportunities to acquire an impressive array of prominent and politically powerful friends. The Irvings came from humble Presbyterian Scottish stock (small-time merchants in the hardware trade), but Washington's numerous elder siblings rapidly ascended in the commercial, intellectual, and political circles of the growing metropolis. The eldest, William, began his career working in the mercantile fur trade with Indians along the Mohawk River. He became an early leader among New York's literati, helping to launch and cowrite Washington's 1807 breakthrough satire *Salmagundi*. William eventually channeled his skill with words to move into politics, representing New York in the U.S. Congress from 1814 to 1819.

Two other brothers received fine educations at Columbia (Peter in medicine and John in moral philosophy). So it was that Washington, at the precocious age of sixteen, was accepted along with his far better qualified brother John to commence legal studies in the offices of prominent Republican attorneys Henry Masterson and Henry Brockholst Livingston. Livingston had been named the state's attorney general after Aaron Burr vacated the post, and his distinguished judicial career took off from there. In 1802 Livingston closed his private practice to serve on the New York Supreme Court,[4] prompting Irving to enter the employment and tutelage of Judge Hoffman. Biographer Andrew Burstein sums it up: "Washington Irving, law clerk, walked a fine line. He liked the society of interesting men. Politics was bringing him in touch with some of New York's most interesting, Federalist *and* Republican."[5] The Irving brothers gravitated to the charismatic Burr, founder of the Society of St. Tammany, which became the center of Democratic-Republican politics. Peter set his medical practice aside to serve as the editor of the *Morning Chronicle*, a newspaper Burr used to instruct his followers and attack his opponents. Through this vehicle, Peter published the first of Washington's Jonathan Oldstyle letters. Washington was establishing his local reputation as a satirist at the same time he was sharing his disdain for the law in that letter to Eaton.

Ebenezer, the third surviving son, stayed relatively clear of politics but contributed by diligently expanding the family's import-export enterprises. Given the volatility of the early national economy during and after the Napoleonic Wars, the whole family depended at critical moments on Ebenezer's ability to keep the business afloat. Sisters Catharine and Ann (Nancy), married and relocated with their husbands to the Mohawk valley, where Washington enjoyed his first encounters with the mystical beauty of upstate New York. Sister Sarah (Sally), closest in age to Washington, would marry a successful businessman and move to a handsome estate in Birmingham, England, where Washington could visit during his extended European sojourns over the subsequent decades. Altogether, Washington's access to society and nature, and the luxury of a career as a creative writer, was entirely facilitated by the support and connections he gained through his brothers and sisters.

Meanwhile, Washington Irving developed a warm social relationship with his employer. In the summer of 1803, Judge Hoffman invited his favored assistant along on a trip to the furthest northern part of New York State and Canada. Hoffman and his cousin Thomas Ludlow Ogden brought their families there to inspect the extensive holdings they had previously acquired through a controversial deal with Holland Land Company. The tracts lay along the Saint Lawrence River, near to where a French mission to the Iroquois had once stood. That settlement of Oswegatchie ("black water") people was now being supplanted by the new town of Ogdensburg. It was an eye-opening and door-opening journey for the imaginative young writer. Irving observed Hoffman's official negotiations, encountered wilderness conditions and Indian groups for the first time in his life, and served as a charming and chivalrous traveling companion to Ogden's and Hoffman's wives and daughters.[6]

After his first trip to Europe (1804–1805), Washington reentered the Hoffman household, gained his admission to the bar in 1806, and served on the Hoffman legal team sent to defend Burr at his federal trial for treason in 1807. He also advanced his reputation as a political satirist with various contributions to *Salmagundi*, another creative project launched by his brother William and William's brother-in-law James Kirk Paulding. Perhaps in token of his sympathy for the embattled former vice president, a youthful Irving jabbed sarcastically in his *Salmagundi* pieces at the president's scientific avocation, reviving that old Federalist campaign ploy while at the same time trivializing the practices of natural history: "The present bashaw is a very plain old gentleman—something they say of a humorist, as he amuses himself with impaling butterflies and pickling tadpoles."[7] During these years Irving's romantic attachment to Judge Hoffman's second daughter, Matilda, grew. Thoughts of marriage and a happy domestic life were dashed, however, when the girl suddenly took ill and died of consumption in April 1809. She was only seventeen. Sublimating his grief into the hard work of writing a more coherent comic masterpiece than the Salmagundi collaboration, Washington published *The Knickerbocker History of New York* later that same year.

With the onset of the war with Great Britain in 1812, Washington Irving forged a lasting friendship with naval hero Stephen Decatur and eventually sought his own glimpse of military life by enlisting for service as aide-de-camp to Governor Daniel Tompkins in 1814. Tompkins remembered Peter and John Irving from their days together as Columbia students. Washington expressed his high regard for the Republican in a letter to brother William: "Tompkins is one of the worthiest men I ever knew. I find him honest, prompt, indefatigable, with a greater stock of practical good sense and ready talent than I had any idea he possessed."[8] Significantly, the one powerful contemporary New York political figure with whom Irving was neither willing nor able to establish cordial relations was De-Witt Clinton. Some of this animosity may have derived from the sympathy the Irving brothers maintained for the disgraced Burr, as well as Washington's close friendship with Samuel Swartwout, the younger brother of Clinton's dueling opposite and key Burr lieutenant John Swartwout. The dislike was mutual; Clinton occasionally carried his grudge against the youngest and most prominent Irving into ungracious public criticism.

Clinton and Irving held fundamentally incompatible attitudes toward scholarly erudition. Whereas the aspiring national political figure sought to manufacture a reputation for serious scientific accomplishment in order to burnish his credentials as a statesman, the upstart humorist regarded such tendentious posturing as a prime target for absurdist ridicule. Yet both men essentially depended on the pervasive cultural relevance of natural scientific knowledge. Mitchill's stuffy but informative 1807 tome *The Picture of New-York* had provided the template for Irving's grand *Knickerbocker History* spoof, published just two years later. Clinton objected, as if he did not get the joke, that Mitchill's history "is pronounced by good judges to contain more solid information than the History of New York by Knickerbocker." The mayor protested in an 1815 pamphlet (pseudonymously published) that the latter work "is really intolerable . . . [for its] heterogeneous and unnatural combination of fiction and history is perfectly disgusting to good taste." Then, showing no forbearance, Clinton issued an explicit ad hominem attack, singling Irving out for his humble mercantile class origins: "Incongruous as the association between the cultivation of ironmongery and literature may appear, yet it is no less true, that another celebrated author, by profession, was originally concerned in a hardware store." Allusions to Irving's informal education and his own low estimate of the man as a conversationalist provided the final straws for Clinton's critique: "As to real science and learning, his mind is a *tabula rasa*: he cannot read any of the classics in their original language; nor does he know the first elements of any science. I have spent an evening in his company, and I find him barren in conversation, and very limited in information."[9]

Diedrich Knickerbocker's narrative clearly violated Clinton's version of what constituted useful knowledge of nature and insulted (for comic effect) the dignity of his friend Mitchill. Irving could not have effectively lampooned Professor Mitchill's pretensions to scholarly comportment, however, had the satirist been ignorant of the terminology and geological ideas of his time. The silly antics and absurd pronouncements of Professor Puddingcoft (Puddinghead) harked back to a time in geological scholarship less than a century removed, when fantastic cosmological systems were advanced by highly respected European theoreticians. The second chapter of the *Knickerbocker History* consists of a remarkably detailed, if tongue in cheek, parade of the cosmogonic proposals that occasioned serious discussion about the Earth's history from ancient times up through the more provocative claims of Buffon, James Hutton, and Erasmus Darwin.[10] Irving prefaced these by alluding to the creation myths of Hindu Brahmins, Mohawk Indians, and Congolese "negro philosophers," ultimately regarding all such accounts as being impossible to disprove and therefore equally authoritative. This kind of anthropological relativism, though it enraged Clinton at the time for undermining science, now seems both quaintly naive and eerily postmodern.

Although it was snobbishly accurate of Clinton to note that Washington Irving had been deprived of the formal university education that so many of his illustrious contemporaries in New York's literary and philosophical clubs had enjoyed, the experiences he gathered while on his grand tour at the age of twenty-one served in some measure as a substitute. Drawing upon evidence from Irving's *Notes and Journal of Travel in Europe*, 1804–05, literary historian William Hedges argues convincingly that Irving worked hard in those years not only to develop an elegant prose style but also to acquire a cultivated man's broad repertoire of ideas, attitudes, social polish, and acquaintance with cultural accomplishment: "He . . . studied landscape with a painter's eye, toured galleries and museums, attended the theater, mixed in polite society, meditated on religious, philosophical, and political questions, and even amused himself with rudimentary scientific experiments—he made observations of volcanic phenomena when he climbed Vesuvius and helped to ascertain the acoustical properties of the cave at Syracuse known as the Ear of Dionysius."[11] If Irving really remained resolutely dumb during that soiree with DeWitt Clinton, perhaps he simply deemed his interlocutor unworthy of entertainment; it would be wrong to assume that the young writer was unprepared to engage the mayor's intellect, even on matters pertaining to the natural sciences.

In his more mature works of fiction, Irving was highly capable of utilizing that landscape painter's eye, if not a finely trained scientific vocabulary, to render vivid observations of the kinds of scenery that contemporary naturalists found so intriguing. For example, when Irving sets the scene for his wildly popular gothic tale, "The Legend of Sleepy Hollow," he reverts at various points to a language that grounds the supernatural anxieties of his characters in conventional poetic portrayals of the natural environment: "Not far from this village, perhaps about two miles, there is a little valley, or rather lap of land among high hills, which is one of the quietest places in the whole world. A small brook glides through it, with just murmur enough to lull one to repose, and the occasional whistle of a quail, or tapping of a woodpecker, is almost the only sound that ever breaks in upon the uniform tranquillity." Later on, as Ichabod Crane approaches the Van Tassel mansion, dreaming opulent dreams of a requited courtship with the voluptuous Katrina, Irving (affecting the resurrected narrative persona of the Knickerbocker) describes the sunset view a traveler might take in from the range of hills that surround the wide Hudson near the Tappan Zee: "A slanting ray lingered on the woody crests of the precipices that overhung some parts of the river, giving greater depth to the dark grey and purple of their rocky sides. A sloop was loitering in the distance, dropping slowly down with the tide, her sail hanging uselessly against the mast, and as the reflection of the sky gleamed along the still water, it seemed as if the vessel was suspended in the air."[12] In both cases, these words enable the reader to suspend a sense of the passage of time. Arresting and calling attention to these normal daylit moments of placidity, Irving stores them up to serve as bucolic points of contrast to the ensuing ghost story told by the antagonist and the mysterious nighttime culmination of Ichabod's misadventure.

Irving would, of course, go on to write influential nonfiction as well. Though his famous biographies of Christopher Columbus and George Washington offered scant opportunities to exercise a geological sensibility, the string of works he produced about the American West after his return from Europe in 1832 incorporated motifs and language that had come to characterize natural history descriptions in the wake of the publication of the Lewis and Clark, the Long, and the Cass-Schoolcraft Expedition reports. Irving's A Tour of the Prairies (1835) exemplified this tradition. Henry Leavitt Ellsworth invited Irving to serve as secretary for a fact-finding mission president Andrew Jackson commissioned in the wake of Black Hawk's rebellion. Ellsworth was appointed to investigate and organize the conditions for Indian resettlement in Oklahoma. Besides the army personnel and expedition guides Ellsworth had been assigned, the company included two European traveling companions Irving had acquired en route during his return to America: Charles Latrobe (an Englishman whose uncle Benjamin was a famous architect and friend of Thomas Jefferson) and the Count de Pourtalès (a young and impetuous Swiss aristocrat). Among his other accomplishments, Latrobe had some acquaintance with natural history: " He was a man of a thousand occupations; a botanist, a geologist, a hunter of beetles and butterflies, a musical amateur, a sketcher of no mean pretensions, in short, a complete virtuoso; added to which, he was a very indefatigable, if not always a very successful, sportsman."[13] This motley crew spent the autumn months of 1832 in the company of Ellsworth's commission, traveling into the Missouri interior from St. Louis through Creek, Delaware, and Osage country, venturing as far southwest as the present-day site of Oklahoma City, and then back to white civilization along the Arkansas River.

Irving had only just obtained his first view of the natural wonder of Niagara Falls when he was introduced to Ellsworth aboard a Lake Erie steamship. After all those years in Europe, and recognizing that Niagara had already become a commercially tamed destination for tourists, Irving was hungry for a deeper taste of wilderness adventure. He would find it in eastern Oklahoma:

> A gallop across the prairies in pursuit of game is by no means so smooth a career as those may imagine, who have only the idea of an open level plain. It is true, the prairies of the hunting ground are not so much entangled with flowering plants and long herbage as the lower prairies, and are principally covered with short buffalo grass; but they are diversified by hill and dale, and where most level, are apt to be cut up by deep rifts and ravines, made by torrents after rains; and which yawning from an even surface are almost like pitfalls in the way of the hunter, checking him suddenly, when in full career, or subjecting him to the risk of limb and life. The plains, too, are beset by burrowing holes of small animals, in which the horse is apt to sink to the fetlock, and throw both himself and his rider. The late rain had covered some parts of the prairie, where the ground was hard, with a thin sheet of water, through which the horse had to splash its way. In other parts there were innumerable shallow hollows, eight or ten feet in diameter, made by the buffaloes, who wallow in sand and mud like swine. These being filled with water, shone like mirrors, so that the horse was continually leaping over them or springing on one side. We had reached, too, a rough part of the prairie, very much broken and cut up; the buffalo, who was

running for life, took no heed to his course, plunging down break-neck ravines, where it was necessary to skirt the borders in search of a safer descent. At length we came to where a winter storm had torn a deep chasm across the whole prairie, leaving open jagged rocks, and forming a long glen bordered by steep crumbling cliffs of mingled stone and clay. Down one of these the buffalo flung himself, half tumbling, half leaping, and then scuttled along the bottom; while I, seeing all further pursuit useless, pulled up, and gazed quietly after him from the border of the cliff, until he disappeared amidst the windings of the ravine.[14]

Being a writer of romances, Irving was compelled to impart a conventional sense of thrill and adventure in his account of the buffalo chase. In addition, however, he interspersed details about the landscape that reveal an outlook well informed by ongoing developments in natural history discourse. By the 1830s, cataclysmic explanations for unusual natural phenomena were no longer considered as scientifically acceptable as they had been two decades earlier. The vast prairies had seemed so strange to woodland-raised European settlers that the first generation of scientific observers who ventured beyond the Appalachian mountains (men like Caleb Atwater) spent a lot of energy invoking tremendous incursions of flood waters or a cyclic pattern of conflagration to account for the general absence of forests in relatively unsettled lands. Somehow, perhaps in his conversations with Latrobe, Irving had instead absorbed the sensibility of Charles Lyell's then-quite-new uniformitarian approach to geology.[15] Though there is no explicit discussion of Earth history in the passage, Irving's description of the prairie highlights the ubiquity of characteristic terrain features, zeroes in on the prominence of what later ecologists would designate as "keystone" species, and attributes even the most dramatic landscape to the action of ordinary causes like animal behavior and periodic rainstorms. In other words, Irving's work presents an interpretation of the prairie as a natural ecosystem whose peculiar geography arose naturally through the interactions of the prevailing climate and the resident organisms. In uniformitarian geology, actual observable processes are shown to be capable of producing all geological phenomena, if given sufficient time to operate. An important corollary effect of adopting uniformitarianism is the implicit acceptance of a potentially vast expansion of time.

After the successful publication of his account of the Ellsworth expedition, Irving took on a couple of historical writing assignments paid for by the fur magnate John Jacob Astor: Astoria (1836) and The Adventures of Captain Bonneville (1837). These works were secondhand narratives wrought from notes and journals recording past explorers' efforts to penetrate the continental interior of North

America. Irving used his talents as a writer to emphasize the dramatic elements of discovery and peril rather than the mundane minutiae of scientific observation. Nevertheless, the literary resonance of geological and botanical research efforts can be examined through these two accounts, which focus more on describing and justifying the outcomes of encounters with warlike aboriginal cultures. *Astoria* tells of the abortive efforts to establish an American trade presence and a permanent settlement near the mouth of the Columbia River in the early 1810s. In *The Adventures of Captain Bonneville*, Irving took up the more recent history of American trappers and explorers in the Rocky Mountains by rewriting and augmenting the personal memoirs of an army officer Astor introduced him to in New York in 1835.

A transcontinental expedition led by William Price Hunt in 1811–1812 takes up the bulk of the chapters in *Astoria*. Notably, though there were two English-born botanists included among the party (John Bradbury and Thomas Nuttall), Irving confines his discussion of their scientific activities to a handful of passages in the 519-page narrative. These few mentions tend to reinforce the older stereotype Knickerbocker presented of natural history's intellectual disconnection from practical concerns:

> Mr. Nuttall seems to have been exclusively devoted to his scientific pursuits. He was a zealous botanist, and all his enthusiasm was awakened at beholding a new world, as it were, opening upon him in the boundless prairies, clad in the vernal and variegated robe of unknown flowers. Whenever the boats landed at meal times, or for any temporary purpose, he would spring on shore, and set out on a hunt for new specimens. Every plant or flower of a rare or unknown species was eagerly seized as a prize. Delighted with the treasures spreading themselves out before him, he went groping and stumbling along among a wilderness of sweets, forgetful of every thing but his immediate pursuit, and had often to be sought after when the boats were about to resume their course. At such times he would be found far off in the prairies, or up to the course of some petty stream, laden with plants of all kinds.[16]

Irving relied here on the insultingly jocular testimony of Nuttall's traveling companion Henry Marie Brackenridge rather than Nuttall's own extensive published report of the geological features he encountered on the expedition,[17] and so *Astoria* unfortunately lacks the observant details Irving had so skillfully supplied to his own more recent expedition account.[18]

Setting aside the dismissive characterization of the expedition's naturalist, the argument of *Astoria* draws upon geological thinking in interesting ways. Overall, the book seeks to advance Astor's reputation as a visionary American colonizer of the Oregon territory, and in so doing, Irving examines rationales for displacing those native peoples who might otherwise obstruct the advance of American civilization. Since not all of the intervening territory was apparently suitable for traditional farming (Long's designation of the Great Plains as the "Great American Desert" was certainly compatible with Irving's account), the question of Indian relocation versus inevitable extermination was a hotly debated political topic in the 1830s. Anticipating the creation of a "Wild West" before any such environment existed in reality or legend, Irving proposes a biological basis for an undesirable social outcome:

> Such is the nature of this immense wilderness of the far West; which apparently defies cultivation, and the habitation of civilized life. Some portions of it along the rivers may partially be subdued by agriculture, others may form vast pastoral tracts, like those of the East; but it is to be feared that a great part of it will form a lawless interval between the abodes of civilized men, like the wastes of the ocean or the deserts of Arabia; and, like them, be subject to the depredations of the marauder. Here may spring up new and mongrel races, like new formations in geology, the amalgamation of the "debris" and "abrasions" of former races, civilized and savage; the remains of broken and almost extinguished tribes; the descendants of wandering hunters and trappers; of fugitives from the Spanish and American frontiers; of adventurers and desperadoes of every class and country, yearly ejected from the bosom of society into the wilderness.[19]

For Irving to use "debris" and "abrasions" as the points of reference for this complicated prognostication suggests not only that he knew something about contemporary geological writing but also that his reading public might find these terms familiar and suggestive. The passage culminates in an impassioned plea against Jackson's forced relocations of eastern Indians. In Irving's opinion, these people would only exacerbate the problem by contributing their shattered tribal remnants to a stew of racially mixed expatriated malcontents.[20]

Irving found in Captain Bonneville's observations from the early 1830s a much richer collection of geological materials to include in the third and final of his nonfiction works on the West. This military man clearly understood the rudiments of topographical and geographical analysis, and so his account conforms to the scientific expedition report genre even more closely than Irving's own tour of the prairies had. For example, the third chapter of Bonneville describes tabular hills in a manner indistinguishable from an early article in *Silliman's Journal*: "The vast plain was studded on the west with innumerable hills of conical shape, such as are seen north of the Arkansas River. These hills have their summits apparently cut off about the same elevation, so as to leave flat surfaces at top. It is conjectured by some that the whole country may originally have been of the altitude of these tabular hills; but through some process of nature may have sunk to its present level; these insulated eminences being protected by broad foundations of solid rock."[21] The standard uniformitarian explanation for this phenomenon proceeded along very similar lines, arguing that an original plain once existed at the level of these summits. This peneplain resulted as the land was eroded by wind, water, and weather over a period of eons, leaving behind the sculpted remains of relatively harder rocks that had once been buried among the softer or more soluble strata. Irving displays cautious approval for this novel explanatory approach when he prefaces it with the phrase, "it is conjectured by some." Based on the following example, however, it is doubtful that Bonneville supplied any awareness of this conjecture himself.

Further west, upon reaching the dramatic scenery of the Snake River valley, the Bonneville narrative lapses into the familiar mode of earlier catastrophist geological description: "The volcanic plain in question forms an area of about sixty miles in diameter, where nothing meets the eye but a desolate and awful waste; where no grass grows nor water runs, and where nothing is to be seen but lava. Ranges of mountains skirt this plain, and, in Captain Bonneville's opinion, were formerly connected, until rent asunder by some convulsion of nature. Far to the east the Three Tetons lift their heads sublimely, and dominate this wide sea of lava — one of the most striking features of a wilderness where everything seems on a scale of stern and simple grandeur." As if to qualify and subdue his protagonist's romantic testimony, and in a virtual echo of Crevecoeur's plea from half a century earlier, Irving immediately adds: "We look forward with impatience for some able geologist to explore this sublime but almost unknown region."[22]

No correspondence remains to suggest that Eaton and Irving kept up their friendship, so it is implausible to suggest that Eaton had any personal influence on the development of Irving's understanding of geology. All that has survived to record an enduring bond between the two men is a story from Joseph Henry's journal recording the topics discussed aboard that first Rensselaer School field trip along the Erie Canal. On a Saturday morning in late May 1826, after having intensively instructed his charges on New York's rock strata for three solid weeks, Professor Eaton kicked back and shared some personal stories: "Mr E related several annecdotes of Wasington Ervin [sic] among others that the origin of his 'Death of a Friend' [a story in The Sketch Book (1819–1820)] was the death of Miss Matilda Hoffman daughter of Mr Hoffman Esq of New York. Ervin was engaged to be married to this young lady when about eighteen years old and she eleven. Miss Hoffman died at the age of eighteen of a consumption & Mr Ervin as ever since remained a batchelor & is now about 43 years old. Mr E was his fellow student in Mr Hoffmans office."[23] While they may never have written or met again over the years, especially given Irving's extensive time abroad, it seems that Eaton retained a lifelong affection for his famous former colleague.

ALL THINGS MUST PASS: WILLIAM CULLEN BRYANT

Washington Irving scholar William Hedges makes a bold claim: "Nothing is at first sight more anomalous in American literature of the early national period than its obsession with ruin and decay. A plethora of graveyard imagery, broken columns, and moss-grown towers undermines assumptions as to what one should encounter in a new country." Though surely illuminated by instances wherein individual lives were cut short prematurely by illness, this is more than the morose dwelling upon the death of, say, a teenaged Matilda Hoffman. Within the context of abstract philosophical Enlightenment debates about history's nature (whether progressive or cyclical), Hedges makes specific claims about two of Irving's literary contemporaries, William Cullen Bryant and James Fenimore Cooper. "Thanatopsis—thoughts of the grave—lie athwart the hope for the future," Hedges remarks, pointing toward both writers' depictions of the prairie as being "but a thin veneer of growth covering the wreckage of unknown civilizations. For them, nature's processional often seems a death rather than a life cycle." Dismissing Bryant's superficial resemblances to William Wordsworth, Hedges critiques him as lacking the English Romantic's "feeling" for nature, instead using "the continental expanses of forest and prairie for graveyard meditation."[24]

Having seen the implications of geologic time insinuate themselves into Irving's mature writings, it seems worthwhile to explore this potential source of insight with respect to both Bryant and Cooper. To my knowledge, historians and literary scholars have not systematically explored the relevance of geology to the question of an early American literary obsession with decay and mortality. A major cultural consequence of taking seriously the scientific study of Earth's history in the early nineteenth century, however, was the new possibility of perceiving the vastness of geologic time. Knickerbocker's caricature of the natural historian as a buffoon might persist even to become a stock character of American literature (the proto-nerd), but the broader implications of deep time were not to be so easily parried by either comic or satiric impulses. The predominance of nature as a favorite subject in Bryant's work is universally recognized, and his lifelong talent for the scientific study of plants is often remarked.[25] But the personal connections to and broader implications of the ideas and concepts he encountered while obtaining his botanical education have not received adequate consideration.

Bryant's earliest published writings might suggest that he viewed the practice of natural history as silly and superfluous rather than a highly valued cultural pursuit. A precocious political satirist brought up among die-hard New England Federalists, at age thirteen "Cullen" (as the boy was known among family and friends) published a witty, much-admired attack on Jefferson's embargo policies. Entering public discourse at virtually the same moment that Irving's *Salmagundi* letters poked fun at the president's hobby of specimen collecting, the rascally teenager from the western hills of Massachusetts was delighting Boston readers by calling for the gentleman farmer to abandon government in favor of his peculiar scientific interests: "Go, search with curious eye, for horned frogs/Mongst the wild wastes of Louisiana bogs/Or where Ohio rolls his turbid stream/Dig for huge bones, thy glory and thy theme."[26] The young Bryant hoped, however, to acquire a classical education. After a disappointing year at Williams College, his aspiration to continue his education at Yale was frustrated when his father was financially unable to support the move. Like Eaton and Irving, Cullen was encouraged to prepare for a law career. He began his legal studies in 1811 in the Berkshires under the direction of Judge Samuel Howe. In 1814 he was shifted to the office of congressman William Baylies, in the eastern part of the state, and he successfully passed the bar and opened his own practice back in the Berkshires the following year.

Though the ivory tower was beyond his reach, the young attorney continued to read widely, and he also began to write poetry, developing his eye for nature description throughout the early years of his legal education and practice. According to Judge Howe's wife Sarah, Bryant was already an avid practical botanist by the time she came to know him in 1813. She recalled how he frequently went "to the woods and fields for his specimens."[27] This avocation had been sparked by the tutelage Cullen received from his father Peter, whose knowledge of medicinal plants allowed him to manufacture his own drugs.[28] By 1819, Bryant had published "The Yellow Violet" and "Green River," which both allude to the beauty and fragility of organisms observed in wild nature. "The Yellow Violet" begins like a cross between an almanac aphorism and an entry in a botanical field guide: "When beechen buds begin to swell,/And woods the bluebirds warble know,/The yellow violet's modest bell/Peeps from last year's leaves below."[29] "Green River" contemplates the dance of light and life at the margin of a mountain stream and recommends the rewards there to be gathered by nature's cognoscenti:

Lonely—save when, by thy rippling tides,
From thicket to thicket the angler glides;
Or the simpler comes with basket and book,
For herbs of power on thy banks to look;
Or haply some dreamer, like me,
To wander, and muse, and gaze on thee."[30]

Bryant's brilliant "Thanatopsis," at least in its initial form, represented the earth merely as a universal multipurpose substrate. As originally published, the poem opened directly with this meditation on an impending death:

Yet a few days, and thee
The all-beholding sun shall see no more
In all his course; nor yet in the cold ground,
Where thy pale form was laid, with many tears,
Nor in the embrace of ocean, shall exist
Thy image. Earth, that nourished thee, shall claim
Thy growth, to be resolved to earth again,
And, lost each human trace, surrendering up
Thine individual being, shalt thou go
To mix forever with the elements,
To be a brother to the insensible rock
And to the sluggish clod, which the rude swain
Turns with his [plow] share, and treads upon. The oak
Shall send his roots abroad, and pierce thy mold."[31]

Resting place for mortal bodies, engine of soil fertility, realm devoid of pain or any other lofty sensibility, this depiction of "nature" conjures neither delightful appreciation nor the appetite for scholarly scrutiny.

The year 1820 was a watershed in Bryant's young adulthood. Hints of ambivalence, such as the wistful final stanza of "Green River," which opens, "Though forced to drudge for the dregs of men/And scrawl strange words with the barbarous pen," signaled publicly what Cullen had shared frequently enough in private correspondence with friends and colleagues: a career in the law promised no more fulfilling a life for Bryant than it had for Irving.[32] Nevertheless, he invested a good five years of his energetic youth building a legal practice in Berkshire County, the part of Massachusetts that borders New York State closest to Albany. On the strength of his emerging professional reputation, he soon was able to begin climbing the rungs of local politics. Bryant's election in February

1819 as Great Barrington's town clerk, his 1820 Independence Day oration (in Stockbridge) attacking the Missouri Compromise, and his appointment later that same year to serve as justice of the peace for Berkshire County, all hinted at an auspicious career in public office. Even as he was enjoying these accomplishments, however, in March 1820 his admired and beloved father Dr. Peter Bryant succumbed to the consumption that had been ailing him for some time. This terrible personal blow came on the heels of other strong reminders that life in early nineteenth-century North America was capricious. Biographer Gilbert Muller argues that a combination of events inspired Bryant to compose "Thanatopsis": the deaths of his grandfather Ebenezer Snell and of the young wife of Dr. Bryant's colleague Jacob Porter in 1813, a typhus epidemic sweeping through the towns of western Massachusetts, his anticipation of his own father's demise, and the Gothic style of England's graveyard poets Robert Blair and Bishop Porteus, whose popular works emphatically impressed the inevitability of death upon their young readers.[33] It is reasonable to conclude that Bryant was sufficiently haunted by all these specters (both real and imagined) to devote himself to the topics of ruin and decay. However, one additional circumstance deserves consideration as a potentially relevant psychological factor shaping Bryant's understanding of the cosmos.

Tramping about the Berkshire hills as he sought to nourish his native botanical curiosity, Bryant found in Amos Eaton a scientific mentor and perhaps also, as Eaton was eighteen years older, a paternal figure. Eaton had known Bryant's father, and Cullen possibly was introduced to Eaton when he was still a child.[34] But the real opportunity to form a bond through their shared interest in natural history came in the spring of 1820. During the first decade after he left his first teaching position at Williams College, Eaton traveled back to New England every year to give public scientific demonstrations. These peripatetic tours frequently brought him to Lenox, Pittsfield, and Great Barrington, where Bryant was becoming a leading citizen. To cultivate the botanical mind of his fiancée, Frances Fairchild, Bryant brought her along on his nature rambles through the woods and mountains of western Massachusetts. The middle-aged Eaton, who believed strongly that talented young women as well as men should take up the pursuit of natural history, was more than happy to guide the young couple.[35] Throughout May and June, Eaton gave occasional public lectures in the evenings, while during the day he made himself available to lead Bryant and Fairchild on several coeducational botanizing excursions in the surrounding hills and forests.

Investigations alongside the aspiring theorist must have added a scientific perspective on the transitoriness of human existence to the many other influences the poet had already imbibed from his reading and life experience. Recalling that Eaton had just completed the second edition of his *Index to the Geology of the Northern States* and that his first opportunity to perform a geological survey for Stephen Van Rensselaer would happen later that same summer, it is not hard to imagine that he would have shared with Bryant and Fairchild his mineralogical observations and conjectures about the rocks they encountered. The Appalachian ranges are very old. Exposed strata in these mountains reveal elongated belts of folded and thrust-faulted marine sedimentary rocks, volcanic rocks, and slivers of ancient ocean floor. Geologists now attribute the birth of these ranges some 480 million years ago to the first of several mountain-building plate collisions that culminated in the construction of a unified supercontinent. Eaton, of course, was innocent of any such plate tectonic interpretations, and it is doubtful whether (as a Wernerian) he would even yet have acknowledged the volcanic origin of the gneiss and granite rocks so common to western Massachusetts. Simply observing, however, the apparent insertion of these harder rocks among dipped and tilted sedimentary rock layers (which themselves would have taken centuries, if not millennia, to be so deposited) might well have induced Eaton and his companions to perceive that time stretched back further than previously suspected.

Over half a century ago, a Harvard-trained literary scholar named Donald Ringe published one isolated article hypothesizing that Bryant's poetry may have been influenced by his knowledge of contemporary geological discourse.[36] Unfortunately, Ringe was unable to reconstruct how and when Bryant might first have become acquainted with that science. The evidence that Ringe did gather produced a time lag between Bryant's apparent awareness of cyclical historical geological dynamics by the very early 1820s and the few exposures to practicing geologists that Ringe could substantiate much later in the poet's life. Noting, for example that Bryant met Charles Lyell while on a visit to England in 1845, Ringe was mystified by the strong geological influences to be found earlier in "The Fountain" (1839) and "A Hymn of the Sea" (1842). He imagined that Bryant's significant contact with geologists must have occurred between 1835 and 1847, while he was editor of the *New York Evening Post*.[37]

Poems Bryant composed or revised immediately after spending a season under Eaton's tutelage, however, suggest that some of the older man's geological ideas profoundly reshaped the poet's understanding of the Earth's history while he was still a comparatively young man. In "A Winter Piece" (composed late in 1820), Bryant's poetry delves for the first time below the Earth's surface. A magical frozen landscape of ice and snow holds our gaze through the first half of the poem until Bryant reveals a glimpse of Earth's hidden mineral treasures:

The spacious cavern of some virgin mine,
Deep in the womb of earth—where the gems grow,
And diamonds put forth radiant rods and bud
With amethyst and topaz—and the place
Lit up, most royally, with the pure beam
That dwells in them."[38]

"The Rivulet" (1823), though it reiterates Bryant's established obsession with the theme of aging and decay, marvels in its celebration of how little the passage of time seems to affect the title object: "Years change thee not. Upon yon hill / The tall old maples, verdant still, / Yet tell, in grandeur of decay. / How swift the years have passed away." Bryant repeatedly congratulates the water's eternal cheerfulness: "The windings of thy silver wave, / And dancing to thy own wild chime / Thou laughest at the lapse of time." Finally, he situates the circle of life and death with respect to the inestimable age of the rivulet:

And I shall sleep—and on thy side,
As ages after ages glide,
Children their early sports shall try,
And pass to hoary age and die.
But thou, unchanged from year to year,
Gayly shalt play and glitter here;
Amid young flowers and tender grass
The endless infancy shall pass;
And, singing down thy narrow glen,
Shalt mock the fading race of men.[39]

Most intriguing of all for my claim that Eaton was a significant influence on Bryant's writing are the modifications he made to his signature work, "Thanatopsis." In 1821, when he published a book collecting poetry that had already appeared in various issues of the *North American Review*, he added a new introduction and conclusion to what was already a widely known and admired poem. The consensus among literary scholars has been to assign Bryant's impulse to make these major alterations entirely to his grief following his father's death.[40] I do not deny the relevance of this major emotional event for Bryant, but, when one reads the poem's revised opening with Amos Eaton as well as Peter Bryant in mind, a fresh and illuminating interpretation becomes possible.

To him who in the love of nature holds
Communion with her visible forms, she speaks

A various language; for his gayer hours
She has a voice of gladness, and a smile
And eloquence of beauty; and she glides
Into his darker musings, with a mild
And healing sympathy that steals away
Their sharpness ere he is aware. When thoughts
Of the last bitter hour come like a blight
Over thy spirit, and sad images
Of the stern agony, and shroud, and pall,
    And breathless darkness, and the narrow house,
Make thee to shudder, and grow sick at heart;—
Go forth, under the open sky, and list
To Nature's teachings, while from all around—
Earth and her waters, and the depths of air—
Comes a still voice. Yet a few days . . . [41]

The poem then continues into its earlier beginning.

Bryant's father may have been the first to instruct him in the ways of communing with botanical secrets, but it was in Eaton's company that Bryant came to hear the differing dialects of Nature's language. In the aftermath of Peter Bryant's death, Cullen went forth under the open sky with his love Frances, and they listened to Professor Eaton discourse upon nature's teachings. Eaton's surviving correspondence provides direct testimony to his early and influential contact with the man who was widely regarded as America's first great poet. On midsummer's eve in 1833, a late night of poetry reading in the Eaton household ended with the aging professor deciding to compose a sentimental epistle to the famous author.

Dear Bryant,

I read your poems from candle-lighting 'till this time [half an hour after midnight], to my wife, wife's sister, daughter Sarah, and two nieces. My 13 year old William (of whose science I boast so much) heard me also. Tears fell like showers at some poems, glee glowed at others, love and friendship softened at others.

Now Bryant (as you and your wife have been my pupils in Botany, and I have your confidence) do tell me plainly. Do these poems come from your little stooping puny self? Or does some singing, pitying & fascinating, angel, use you as his vehicle, to relieve cold, cloddy man from the harsh tones of hardy science? I am frightened at the thought of having been your teacher in 1820. Did you know then, that you was destined to charm your clamorous coarse old teacher? Did you laugh when I affected to be your superior, because I knew the names of more weeds than you? Had the sacred nine then called on you? Did you translate, in fancy, my Claytonia, my Anemone, my Solidago, etc. into verse; which you now sing so charmingly? Did you then destine me to early Death, whose terrors you have almost annihilated in your consoling hymn? Did you think of me, when you made your guide-board to the woods? Did Green River (whose banks I trode with you) call up your remembrance of your old schoolmaster? There I shewed you the Wind-flower, and traced its tender organs.

Tell me plainly—is a poet truly a Vâtes? Did you really feel your heavenly birth, when I gave you the name of calyx, corol, and stamen, with loftily affected look?

Your inspired old Schoolmaster, Amos Eaton[42]

A polite but sincere reply quickly came. Bryant declined to address any of the rhetorical questions Eaton had posed (for the record, the poem "Green River" was committed to paper well before Eaton's guided natural history walk visited the banks of its subject), but he thanked the professor for his fine compliments and closed the note by acknowledging: "For the guidance in my botanical studies to which you allude, I have ever held myself your debtor; and that you may long live to diffuse a taste for the sciences you pursue with so much ardor and success, is the prayer of Your sincere friend, William C. Bryant"[43]

By 1833, of course, Bryant had arrived in a place very different from what Eaton might have anticipated when they parted company thirteen years earlier. Bryant's discontent with a career in law (something else he and Eaton had in common) proved impossible to ignore, and after a few exploratory trips to New York City, the thirty-one-year-old writer took the great leap and moved his young family to the burgeoning metropolis in 1825. Arriving just in time to witness the ceremonies officially opening the Erie Canal to commercial traffic, Bryant managed the extraordinary feat of making his living with his pen. Steady production of poetry, his editorship of the *Evening Post*, and his active membership in the leading literary clubs brought Bryant into contact with not only the Empire State's leading poets, playwrights, and novelists but also the men and women deeply interested in the practices of natural philosophy and natural history. The civic organizations DeWitt Clinton had founded with such energy and purpose now became familiar haunts for Bryant and his new friends. But where Irving had found it necessary and attractive to remove for years on end to England and Europe, Bryant pressed ahead to develop a self-confident American literary voice. In the classic appraisal of mid-twentieth-century literary historian Van Wyck Brooks, Bryant "was the first American poet who was wholly sympathetic with the atmosphere and feeling in the country and who expressed its inner moods and reflected the landscape, the woods and the fields as if America itself were speaking through him."[44]

When he received Eaton's delightful letter, Bryant was fully engaged in this project of using American nature to erect an independent literature that would be respectable according to the models provided by Europe's great writers. Eaton, of course, had done the same with regard to American science. "The Prairies" (1833) offers an illustrative example of what form this challenge took in the realm of poetry. It begins: "These are the gardens of the Desert, these / The unshorn fields, boundless and beautiful, / For which the speech of England has no name— / The prairies." As Walt Whitman and other great American poets would each try in turn, Bryant sought to convey his sense that the scale of nature in the New World was exaggerated, as compared with that of Europe. Thus, even the English language somehow had to be stretched beyond its Old World sensibilities to accommodate the challenge of describing American natural phenomena.

> I behold them for the first,
> And my heart swells, while the dilated sight
> Takes in the encircling vastness. Lo! They stretch
> In airy undulations, far away,
> As if the Ocean, in his gentlest swell,
> Stood still, with all his rounded billows fixed,
> And motionless forever.

To gain some purchase on the awesomeness of the scene, natural history and human history blend as Bryant contemplates his perennial theme of death:

> Let the mighty mounds
> That overlook the rivers, or that rise
> In the dim forest crowded with old oaks,
> Answer. A race, that long has passed away,
> Built them; a disciplined and populous race
> Heaped, with long toil, the earth, while yet the Greek
> Was hewing the Pentilicus to forms
> Of symmetry, and rearing on its rock
> The glittering Parthenon.

In line with contemporary scientific discourse about the origin of the Western "antiquities," Bryant interjects in his poem the rationale Jacksonians had adopted to justify the dispossession of Indian tribes as savage interlopers who must have extinguished the cultures that had built such wondrous ancient earthworks:

The red man came—
The roaming hunter-tribes, warlike and fierce,
And the mound-builders vanished from the earth.
The solitude of centuries untold
Has settled where they dwelt. . . .
All is gone;
All—save the piles of earth that held their bones,
The platforms where they worshipped unknown gods,
The barriers which they builded from the soil
To keep the foe at bay—till over the walls
The wild beleaguerers broke, and, one by one,
The strongholds of the plain were forced, and heaped
With corpses.[45]

Literary analyses of Bryant's post-1835 works take up the argument about how use of geological imagery may have grown and changed over the course of his lifetime. Ringe points us to passages where dynamic theories of the Earth clearly informed the poet.

Haply shall these green hills
Sink, with the lapse of years, into the gulf
Of ocean waters, and thy source be lost
Amidst the bitter brine? Or shall they rise,
Upheaved in broken cliffs and airy peaks.
Haunts of the eagle and the snake, and thou
Gush midway from the bare and barren steep?[46]

The other example Ringe cites stresses the divine Creator's hand in using erosion and gradual uplift (two key engines of Lyell's uniformitarian geology) to shape and reshape the world.

These restless surges eat away the shores
Of earth's old continents; the fertile plain
Welters in shallows, headlands crumble down,
And the tide drifts the sea-sand in the streets
Of the drowned city. Thou, meanwhile, afar
In the green chambers of the middle sea,
Where broadest spread the waters and the line
Sinks deepest, while no eye beholds thy work,
Creator! thou dost teach the coral-worm

To lay his mighty reefs. From age to age,
  He builds beneath the waters, till, at last,
His bulwarks overtop the brine, and check
The long wave rolling from the southern pole
To break upon Japan. Thou bidd'st the fires,
That smoulder under ocean, heave on high
The new-made mountains, and uplift their peaks,
A place of refuge for the storm-driven bird.[47]

In light of my contention that Eaton may have provided Bryant with a geological education over a decade earlier, Ringe's conclusion that contemporary geology provided only imagery and evidence for Bryant but not a defining worldview must be revised. He notes, correctly, that Bryant's cyclical view of history was already well established by 1821, when the poet recited his lengthiest work, "The Ages," composed at the invitation of the Phi Beta Kappa society to be delivered at Harvard's College commencement exercises in Cambridge that spring.[48] If, however, Eaton exposed Bryant and his fiancée to startling evidence of "deep time" in 1820, this geological perspective may have helped the young poet to formulate the sophisticated philosophical outlook in his contemplation of eternity. Geology was the science that actively interrogated "timeless" natural phenomena, seeking a more detailed understanding of dynamic processes that could create or destroy even the most imposing landforms. After Bryant, American literature would no longer be untouched by that geological suspicion of humanity's temporal insignificance.

LEATHER-STOCKING PHILOSOPHER: JAMES FENIMORE COOPER

James Fenimore Cooper was born in 1789, thirteen years after Eaton, six years after Irving, and five years before Bryant. His class background was no more distinguished than theirs (whose fathers were, respectively, a farmer, a merchant, and a medical doctor), but Cooper's father sought to bestow social privilege and material advantages to his children. (Like Irving, Cooper had many elder siblings.) A particularly adroit land speculator of Pennsylvania Quaker heritage, William Cooper worked ambitiously in the decade after the Revolutionary War to set himself up as a "gentleman squire," amassing an estate from scratch in central New York State that resembled a patroonship in its size and political influence. He was particularly assiduous in courting the favor of New York Federalist leaders; Josiah Ogden Hoffman and Stephen Van Rensselaer numbered among Cooper's numerous investment partners and patrons.[49] The successful launching of Cooperstown, however, depended not only on cultivating prestigious connections but also on recruiting and retaining committed settlers from among the land-hungry veterans (especially hardy New Englanders) who were willing first to subdue the forested uplands that surrounded Lake Otsego and then to transform their holdings into productive farms.

This region, which James Fenimore Cooper fictionalized in his Leather-Stocking novels (so named for the garb associated with their shared protagonist, Natty Bumppo), would become enshrined in early American literature as indelibly as Irving's legendary, Dutchified Hudson Valley and Catskill Mountains. As a child, Cooper explored those woods and hills, absorbing nature lore from the trappers and hunters he encountered. Ultimately, he developed a philosophy that blended the contrasting impulses of his upbringing. His father's energetic self-promotion inspired Cooper with an appreciation for individual self-reliance but raised concerns about personal morality and social class stability that would resonate throughout his novels. Most importantly, Cooper became a steadfast champion of the kind of wisdom and judgment that derives from extended experience in a natural wilderness environment, and this ineffable quasi-spiritual quality is what distinguishes the Leather-Stocking philosopher from possessors of other kinds of nature knowledge (whether scientific in a learned or a practical manner). The balance of this section is devoted to an examination of the relationship of early American republic natural history knowledge enterprises to Cooper's intellectual life and work, especially as represented (for purposes of comparison and contrast to Irving and Bryant) in the final fictional installment of Natty Bumppo's career, *The Prairie* (1827).

Cooper enjoyed an unusual education, the study of which provides useful hints that can help us to understand the complicated attitude he would display toward natural history in his creative works. William Cooper's efforts at social elevation were somewhat hampered by his incorrigibly crude manners, but this shortcoming did not prevent him from mobilizing his political connections and rapidly growing material wealth in a bid to purchase aristocratic class identity for his children. He briefly enrolled his youngest son James at the common school he had provided for Cooperstown's rustic settlers; then in 1801 he remanded the rambunctious twelve-year old boy to the care of a family friend, an Oxford-educated Episcopal clergyman who operated a small school in Albany. In the Reverend Ellison's elite classroom, young Cooper was tutored alongside scions of leading Federalist families. There Cooper formed a lifelong bond of friendship with John Jay's son William. Thus admitted to the governor's familial and social circles,

Cooper soon became intimately acquainted with several other privileged Albany boys his age, including the Patroon's son Stephen Van Rensselaer IV.[50]

When illness cut short their Albany schoolmaster's life, James was sent along with his brother Samuel to New Haven to continue his formal education. With the assistance of a private tutor, James spent the fall of 1802 obtaining a rudimentary reading knowledge of Greek to add to his well-established skills in Latin and arithmetic. Though not particularly passionate about his studies, James soon passed the admission exam and entered Yale as a thirteen-year-old college freshman in February 1803. This was a fortuitous moment in the history of American science, for President Timothy Dwight had just appointed the twenty-one-year-old Benjamin Silliman to become the college's first professor of chemistry and natural history. The man who would later serve as Eaton's mentor in mineralogy and geology was at this time a total scientific neophyte. After a brief sojourn to Philadelphia to gain some know-how and purchase basic laboratory equipment, Silliman offered his first set of scientific lectures at New Haven in the spring of 1804. He hired Cooper, then a sophomore, to serve as his assistant in running experiments and lecture demonstrations for that year and the next. Biographer Wayne Franklin doubts that Cooper learned much natural scientific knowledge, but the time spent under Silliman's tutelage would have given him a direct "insight into the scientific method" and probably established the intellectual basis for Cooper's later friendship and collegiality with New York naturalists James Renwick and James De Kay.[51]

What Franklin does not emphasize is the opportunity Cooper had to incorporate the quest for scientific knowledge about natural things into his own emerging worldview. Working with the inexperienced Silliman, Cooper was examining and cataloging the beginnings of Yale's collection of botanical and mineral specimens. Growing up in Cooperstown, Cooper likely developed curiosity about the science of mineralogy. As historian Alan Taylor notes, natural products that were easily harvested and offered the highest value per weight (such as gemstones and exposed mineral veins) were especially alluring to entrepreneurs in frontier economies: "[William] Cooper took a special interest in rumored ores and mines, an interest greatly disproportionate to the actual prospects of finding rare minerals in upstate New York. . . . Driven by wishful thinking, he recurrently collected specimens of purported ores—copper, iron, and silver—for eager dispatch to experts in the cities. Alas, all of Cooper's samples proved of little or no value."[52] So, it is hardly a stretch to suggest that Cooper's experiences, at home and with Silliman, enabled the future novelist to recognize contemporary scientific practice as one of the significant modes of thinking available to any person trying to make sense of the natural world.

By the winter of 1805, the deficiencies of Silliman's scientific education had become all too clear, and so he prepared to make an extended visit to Europe. For the next twelve months, he attended lectures on geology at the University of Edinburgh given by the leading Wernerian and Huttonian theorists of his day. He also made connections with other colleagues in Holland, France, and England and spent the $10,000 the fellows of Yale College had appropriated to purchase books for the library and state-of-the-art European "philosophical and chemical apparatus."[53] Silliman returned in the spring of 1806 finally equipped to live up to his job title.

Cooper's departure from Yale was neither so well planned nor so auspicious. A series of violent altercations between Cooper and some southern students led to a bloody beating in May 1805 that left Cooper's face and head sufficiently injured that the New Haven city judicial system became involved. The following month he was expelled from Yale for setting the dormitory door of one of his assailants on fire (perhaps using knowledge gained through his work with chemical explosives ingredients).[54] Determining that something more active was required for the troubled teenager, William Cooper acquiesced to James's request that he be allowed to go to sea. Spending his next two years aboard a merchant marine vessel, Cooper made his first journey to the Mediterranean. He returned home an experienced seaman, and with the declaration of the Embargo Act in 1807, decided to join the U.S. Navy. His subsequent assignment as a midshipman aboard the U.S. brig *Oneida* stationed Cooper on New York's Lake Ontario shore, affording his first opportunity to visit and absorb the natural wonder of Niagara Falls and to witness the tense build-up to the War of 1812 from within the culture of the armed services.[55]

By 1825 Cooper had married and was renting one of the Patroon's fine New York mansions (directly across the street from fur magnate John Jacob Astor) and working as a successful writer. He had gained national celebrity after the publication of his second novel *The Spy* (1821), and he used this fame to command the attention of leading intellects who had expressed interest in advancing the cause of American literature. A flourishing network rapidly sprouted up around Cooper, as he invited aspiring writers, artists, and natural scientists to interact under his auspices with New York's publishers and prominent economic and political leaders. A private group, dubbed the Bread and Cheese club, supplemented the panoply of public institutions that DeWitt Clinton had worked so hard to establish a decade earlier, such as the New York Lyceum, the Literary and Philosophical

Society, and the New York Historical Society (all of which Cooper and his contemporaries patronized and enlivened). In practical terms, the Bread and Cheese
Club amounted to a biweekly lunch meeting where a wide variety of intellectuals and their patrons could exchange creative ideas. The original Knickerbocker,
Washington Irving, though living abroad at the time, was named honorary chair-
man of the club in 1824, but Cooper was unquestionably the convener and leader
of this important mid-1820s New York cultural institution. By 1825, when William
Cullen Bryant arrived on the New York literary scene, the thirty or so regular
attendees of the Bread and Cheese club included: Knickerbocker writers and
editors Fitz-Greene Halleck, James Kirke Paulding, and Gulian C. Verplanck;
the publisher Charles Wiley; landscape painters Thomas Cole, William Dunlap,
Asher B. Durand, Henry Inman, Samuel F. B. Morse, and John Wesley Jarvis;
prominent New York scientists David Hosack, James De Kay (Stephen Long's
expedition naturalist), and New York Academy of Medicine president John Wake-
field Francis; Columbia law professor James Kent; and Philip Hone, a merchant
and president of the Delaware and Hudson Canal Company who, though a political Whig during a time when Jacksonian Democrats were gaining immense
popularity in the city, was elected in 1826 to serve one term as New York's mayor.[56]
When the English scientist John Finch first arrived in New York in 1824, it was
natural enough for his local host, Dr. Hosack, to bring him along to meet all of
these luminaries at "the Lunch." Finch was surprised to discover that the toast of
New York's literary crowd traveled in the same circles as the scientists.[57] In his recollections of that illustrious occasion, Finch noted that Bryant and Cooper were
as keenly interested in the geological sciences as were many of their Romantic
counterparts on the other side of the Atlantic—William Wordsworth and William
Blake, the Shelleys (Mary Wollstonecraft and Percy Bysshe), and Lord Byron.[58]

    Although no documentation exists to prove that Cooper ever met Amos
Eaton,[59] he did know Silliman and Van Rensselaer, the men whose knowledge
and resources effectively launched Eaton's productive career in science. Cooper
and Eaton therefore lived within overlapping temporal, geographic, and social
milieux. The interesting historical point to investigate here is whether Cooper's
literature may also have drawn substantially upon an ambient cultural embrace
of geology, even in the absence of the kinds of specific encounters that Bryant and
Irving experienced with Amos Eaton.[60]

    Ironically, writings about the trans-Mississippi West provide the simplest il-
lustrations for a comparative analytical framework for studying this triumvirate of
New York's early literary giants. Half a dozen years before either Irving or Bryant
would venture to write anything about the Great Plains, Cooper completed a

third installment of his Leather-Stocking tales to follow *The Pioneers* (1823) and *The Last of the Mohicans* (1826). In *The Prairie* (1827), Cooper crystallizes the American archetype of a ridiculous scientific antithesis to the natural wisdom of the Leather-Stocking hero. Unlike the wilderness-wise trapper (an aged Natty Bumppo), the character Obed Bat (a.k.a. Doctor Battius) is a formally educated man. He is an expert in all aspects of European natural history practices. With intentional irony, Cooper lampoons the zoological uses of the scientific rules of binomial Latin nomenclature. Absurd episodes, such as when Battius mistakes his own donkey in the darkness for an indigenous batlike(!) species of quadruped *Vespertillio Horribilis, Americanus*, punctuate the novel's plot with comic relief.[61]

In the frontier world of the novelist's imagination, fine intellectual accomplishments and scientific virtues amount to hardly anything worthwhile, since the quest for ultimate natural knowledge seems to transport the seeker to some ethereal place rather than any real plane of observation. The narrator comes right out and says as much at one point: "The worthy naturalist belonged to that species of discoverers who make the worst possible travelling companions to a man who had a reason to be in a hurry. No stone, no bush, no plant, is ever suffered to escape the examination of their vigilant eyes, and thunder may matter, and rain fall, without disturbing the abstraction of their reveries."[62] Though Obed Bat can be comically timid in the presence of unfamiliar *horribilis* creatures, in his enthusiasm for discovery he is likely to recklessly shed all sense of precaution. At one moment, while the group is pinned against a rocky ledge by some armed Indians, "the eye of the naturalist had caught a glimpse of an unknown plant, a few yards above his head, and in a situation more than commonly exposed to the missiles which the girls were unceasingly hurling in the direction of the assailants. Forgetting, in an instant, everything but the glory of being the first to give this jewel to the catalogues of science, he sprang upward at the prize with the avidity with which the sparrow darts upon the butterfly."[63]

Though not nearly so harsh as Irving's Knickerbocker skewering of Samuel Latham Mitchill's pretensions, the presence of the educated naturalist in the American West gives Cooper a chance to catalogue and morally distinguish ways of knowing nature. By the time he reaches the prairie, Natty Bumppo has acquired a lifetime of wilderness know-how, gathered while passing through a variety of natural landscapes and in the close company of indigenous people rather than just European settlers. Paul Hover, another character in the novel, possesses the practical skills of an intelligent agriculturalist, as typified by his frequently used sobriquet, "the beekeeper." And then, of course, there is Obed Bat. Wayne Franklin suggests that this caricature of the natural scientist may derive in part

from Cooper's early experiences working closely with Benjamin Silliman.[64] For
the character's penchant to let his imagination run away with the discovery and
naming of new species, however, one real life model to consider is Constantine
Rafinesque, the foreign-born naturalist whose notoriety for publishing new species and genera names by the dozens and hundreds was already the cause of such
controversy in American botany and zoology in the 1820s.[65] Most intriguing of all,
however, is the manner in which Cooper assigns specific exaggerated behaviors
to Doctor Battius—behaviors that resemble the eccentricities Astoria expedition
members reported about traveling with the botanist and ornithologist Thomas
Nuttall. Cooper scholar Robert Madison confirms this identification and goes on
to summarize nicely: "Obed Bat really has two roles in the first half of the novel:
he is the idiot savant . . . but he is also, much more importantly, I think, the ever-
present reminder that *Prairie* is 'about' breeding. Obed Bat pays *constant* attention
to classification—he is in many more action scenes than one would expect, but
is always distracted by some rare breed of plant or animal."[66]

Cooper's treatment of natural scientists shares important elements with Irving's
invented histories. But to what degree do his works show evidence of a substantial
embrace of contemporary geological thinking itself? One Cooper scholar insists
that the depiction of accurate geological knowledge was indeed a specific concern of the novelist in 1831 when he began to compose his European sea tale, *The
Bravo*: "Cooper worked hard in manuscript with the passages on Venetian geology and its unique canals."[67] This active interest in geology, which Finch claimed
was already in Cooper's conversation in 1824, would be powerfully invoked in *The
Prairie*. Surprisingly, Cooper wrote the novel while residing in Europe, though
he had never actually seen prairies for himself. As his daughter Susan Fenimore
Cooper would later recall: "His travels westward had not extended farther than
Buffalo and Niagara, where he had gone on duty, when serving in the navy. And
at the moment of planning [*The Prairie*], he had not leisure for an excursion
beyond the Mississippi, much as he wished to see that singular region. The necessary information could, therefore, be drawn from books and conversation only."[68]

The vicarious nature of Cooper's encounters with distant places may actually have enhanced the hunger he displayed for staying abreast of current geological debates. While staying in Paris in June 1832, Cooper composed a new
introduction for a revised edition of *The Prairie* that begins with two paragraphs
about historical geology. The technical language Cooper adopts here could easily
have graced a contemporary article in *Silliman's Journal*, whereas the theoretical
thrusts were unlikely to have been heard yet outside of a few European scientific
capitals:

The geological formation of that portion of the American Union, which lies between the Alleghanies and the Rocky Mountains, has given rise to many ingenious theories. Virtually, the whole of this immense region is a plain. For a distance extending nearly fifteen hundred miles east and west, and six hundred miles north and south, there is scarcely an elevation worthy to be called a mountain. Even hills are not common; though a good deal of the face of the country has more or less of that "rolling" character, which is described in the opening pages of this work.

There is much reason to believe that the territory which now composes Ohio, Illinois, Indiana, Michigan, and a large portion of the country west of the Mississippi, lay formerly under water. The soil of all the former States has the appearance of an alluvial deposit; and isolated rocks have been found, of a nature and in situations which render it difficult to refute the opinion that they have been transferred to their present beds by floating ice. This theory assumes that the Great Lakes were the deep pools of one immense body of fresh water, which lay too low to be drained by the irruption that laid bare the land.[69]

The contents of this passage are close to the cutting edge of contemporary scientific discourse, given how early these words were written. To have asserted the agency of floating ice for erratic boulders in 1832 (rather than the kind of diluvial explanations William Buckland had proposed and Eaton had elaborated for North America throughout the 1820s), Cooper had to have been in contact with people who were debating Charles Lyell's ideas on floating ice before they were published in volume 3 of his *Principles of Geology* in May 1833.[70] Cooper's embrace of the cyclical nature of uniformitarian geology would eventually bring him close to the views Bryant had developed independently of any reading of Lyell. As American literature scholar Charles Adams observes, with respect to Cooper's late novel *The Crater* (1847), "Just after the earthquake has created Mark's second 'New World,' Cooper paraphrases his scientific reading to explain the geological processes which could bring about such an event." Adams concludes his trenchant analysis of the role scientific ideas play in the novel by noting that: "Geology affords Cooper a lexicon by which to articulate his deepest convictions about the relationship between the individual life and the structures—natural and human—that condition it. Like the law elsewhere in his canon, science in *The Crater* becomes a means of imagining the world; a metaphoric structure describing the 'uniformity of state' that Cooper sought to establish in his fiction, however he might despair of finding it in the world."[71]

In addition to Cooper's fiction, his nonfiction can also be examined to see whether and how, as was the case with Irving, Cooper's awareness of geological ideas directly penetrated any of his social or political criticism. *Notions of the Americans* (1828) stands out as Cooper's major contribution to that same genre DeWitt Clinton had graced (as Hibernicus) with his *Letters on the Natural History of New-York* (opinionated observations of American society offered by a native author, but under the guise of being a foreign visitor). Interestingly, no extensive application of geological reasoning figures in this work. Written during Cooper's extended residence in Europe, the book must have relied upon notes and recollections from his journeys around New York and New England throughout the previous two decades. Perhaps that time lag explains why Cooper portrays the Hudson Highlands in the same kind of scenic language and dramatic analysis that writers like Samuel Mitchill and William Darby used in the burgeoning travel literature that stimulated the popular impulse to tour remote places in New York State starting in the 1810s. Lacking the Knickerbocker travel writer's tone of ironic perspective, Cooper self-consciously embraces the visually dramatic in this mode of heroic historical geological description: "These [highlands] are a succession of confused and beautifully romantic mountains, with broken and irregular summits, which Nature had apparently once opposed to the passage of the water. The elements, most probably assisted by some violent convulsion of the crust of the earth, triumphed, and the river has wrought for itself a sinuous channel through the maze of hills, for a distance of not less than twenty miles."[72] Cooper's narrator, adopting the persona of a European visitor, consistently employs an aesthetic rather than a scientific or political philosophical basis for his social and moral criticism. So, for example, after comparing American to European mountains his general conclusion is that the former are easier to till because they are "less abrupt." According to the visitor's facile logic, whatever American mountains may lack in their inferior beauty, they make up for with their greater utility.[73]

It is a commonplace to observe that Washington Irving, William Cullen Bryant, and James Fenimore Cooper were all engaged in the collective task of inventing a "history" for the new nation.[74] Irving's approach to the challenge of historical fabrication was, in William Hedges's fine turn of phrase, "to equip the American landscape with a kind of mythology, to crystallize a handful of memories, traditions, and a connection with a past already blurred and threatened with extinction."[75] Bryant invoked a clearer and more immediately scientific grasp of nature in his solution to the same problem. Van Wyck Brooks distinguishes the Berkshire poet's fresh approach in this manner: "Bryant was a botanist who had studied Linnaeus as closely as the Bible and Homer, and the delicate descriptive touches in his flower-pieces were drawn from exact observation and definite knowledge."[76] In his *Notions of Americans*, Cooper railed against all the creative disadvantages facing the American writer, who could rely upon "no annals for the historian; no follies (beyond the most vulgar and commonplace) for the satirist; no manners for the dramatist; no obscure fictions for the writer of romance; no gross and hardy offenses against decorum for the moralist; nor any of the rich artificial auxiliaries of poetry."[77] Notwithstanding all these handicaps, Cooper was able in his own fictional works to immortalize a remarkable cast of original American characters and vivid locations. As one early twentieth-century literary critic mused: "a clumsy stylist, a prosy poet, he yet contrives to place before us this grand scene—the breathing stillness of the ancient woods, whose oaks and pines, free of underbrush at the base, soar into the upper air; unsullied lakes reflecting the image of fair islands and mountains and the moods of the sky; the laugh of the loon, the tapping of woodpeckers, the sound of waterfalls."[78]

All three writers found the science of geology—both in its particular applications to the study of New York's landscape and in its broader philosophical implications for an appreciation for the expansiveness of Earth's history—to be a provocative and worthy source of intellectual inspiration. Writers seeking tools for the invention of history, in a land where written archives of human chronicles were relatively skimpy, understood that only two other alternatives beckoned: the geological history that was inscribed into the land itself and the memory of the indigenous people who predated European immigration. A lot of interesting literary scholarship has been invested in analyzing the latter of these sources of inspiration; relatively less ink has been spilled on the former.[79] The advice Crevecoeur originally offered, about discerning the answer to "What is an American?" by investigating the nature of the land itself, was a lesson New York's pioneering writers of stories, poems, and novels evidently took to heart.[80]

CHAPTER EIGHT

Kindred Spirits

This is said of the poet; but the landscape painter is admitted to a
closer familiarity with nature than the poet. He studies her aspect
more minutely and watches with a more affectionate attention its
varied expressions. Not one of her forms is lost upon him; not a
gleam of sunshine penetrates her green recesses; not a cloud casts
its shadow unobserved by him; every tint of the morning or the
evening, of the gray or the golden noon, of the near or the remote
object is noted by his eye and copied by his pencil. All her boundless variety of outlines and shades become almost a part of his being
and are blended with his mind.

—*William Cullen Bryant, 1848*

HUDSON RIVER SCHOOL LANDSCAPE ARTISTS

While geology's relevance to early American literature may have been scarcely appreciated by previous scholars, the same cannot be said with reference to studies
of Romantic landscape art.[1] Art historian Barbara Novak led the way thirty years
ago when she noted how Americans in the early republic who were grappling
with the country's cultural inferiority complex (vis-à-vis Europe) were delighted
when Hudson River school landscape artists took particular advantage of the real-
ization that "with every geological discovery, America got older."[2] Rebecca Bedell
has expanded on this insight in her marvelous book *The Anatomy of Nature*.[3] Kenneth Haltman's incisive study of the watercolors by British-born Philadelphian
Samuel Seymour, who was the official painter for Maj. Stephen Long's journey
from Pittsburgh to the Rocky Mountains in 1819 and 1820, also shows how geological knowledge and theoretical implications could be interdependently conveyed
through the images and narrative output of a scientific expedition.[4] Early American landscape artists engaged seriously with the science of geology, and members
of the Hudson River school were in an especially advantageous position to apply

this new science to their depictions of New York's many picturesque mountains, lakes, and waterfalls.

Where Washington Irving and James Fenimore Cooper inscribed scenic legends and an original sense of history onto the Hudson valley, Lake George, and the Catskill Mountains with the written word, painters such as Thomas Cole, Asher Durand, and Frederic Church visually captured scenes of natural beauty and spiritual power from these and wilder places (such as the Adirondack Mountains and the Niagara and Genesee River cataracts). These artists and their colleagues would explore similar themes of wild and sublime nature in places ranging from New Hampshire's White Mountains to the majestic rivers and Rocky Mountains of the American West, and as far away as the Amazon rainforest and Andes volcanoes in South America. Along the way, Hudson River school painters developed an entirely new visual vocabulary to represent the distinctive experiences of American nature.

William Cullen Bryant's close personal friend, the English-born painter Thomas Cole, is generally regarded as the founder and leading exemplar of the Hudson River school. At first glance, Cole's works stand out as realistic visual representations of iconic American places. Before delving into the question of how natural history practices supported the "photographic" quality of these paintings, however, one might note that Thomas Cole was a painter of epic historical ambitions. Drawing heavily from literary and historical sources, Cole expressed deeply held religious and philosophical convictions through allegorical interpretation.[5] He shared Bryant's cyclical view of history (natural and human), and his artistic achievements probably exceeded the poet's in their impact. Asher Durand's famous painting *Kindred Spirits* commemorated the friendship between Bryant and Cole by picturing the two men apparently conversing after a hike to the top of a rock overhanging the gorge below the lower drop of Kaaterskill Falls, a location that had become a popular New York literary tourist destination soon after Irving published "Rip Van Winkle" three decades earlier.[6] Significantly, Durand's picture celebrates the spiritual bond between the then-still-living poet and the recently deceased artist (Cole died suddenly of pneumonia in 1848) in the midst of this famous natural place. As was common in many European and American early nineteenth-century landscapes, the foreground of the composition features scraggly trees, decaying branches, and dead stumps. The inclusion of these features might specifically evoke Cole's own (1825) study entitled *Lake with Dead Trees (Catskill)*, one of a group of three paintings that had launched the self-taught Cole's career by attracting the attention of leading New York painters and art collectors William Dunlap, John Trumbull, and Durand himself.[7] Stumps conventionally told of the effect of human encroachment into a wilderness, just as decay always implied aging and the lapse of time. However, Durand's painting displays awareness of a much deeper record of time's passage. Across from the pair of friends, the eroded but imposing mass of the facing cliff, with its finely delineated horizontal layers of sedimentary rock, attest to the dramatic power of this small stream's erosive capability when applied over countless millennia.

Bryant's funeral oration for Cole (part of which is quoted as this chapter's epigraph) hints at his extraordinary contribution to the working lexicon of American art. Drawing on prolonged scientific study and extensive field practice, Cole introduced into American art the representation of geological features and their Earth history implications. Cole became an exceptionally devoted student of natural history. From the start, he attracted companions and patrons who could nurture his interest in geology. The English-born geologist George William Featherstonhaugh, for example, took an interest in Cole very early in his career. Although things ended badly between the two men,[8] an invitation in 1825 to paint Featherstonhaugh's estates in Duanesburg, New York, gave the artist entry into Albany's scientific circles, where he met Stephen van Rensselaer (and, though it is unconfirmed, very possibly Amos Eaton).[9] Columbia's professor of botany David Hosack would purchase, on fairer terms, three of Cole's most interesting early literary/historical paintings: The Last of the Mohicans (1827), Expulsion from the Garden of Eden (1828), and The Subsiding of the Waters of the Deluge (1830).[10]

Cole was an active member of James Fenimore Cooper's Bread and Cheese club. Besides becoming a very close friend of William Cullen Bryant and the scientifically minded artist Samuel F. B. Morse, he would also have found regular opportunities to converse with Dr. Hosack and the city's other leading naturalists. Outside of the city, Cole cultivated Benjamin Silliman's brother-in-law Daniel Wadsworth to be another one of his art patrons, a move that gained him social access to the New England scientific community. This relationship also provided a conduit for Cole to receive reliable information about new scientific developments.[11] He joined Morse and Wadsworth on several fossil collecting and geological sketching field trips to various places in upstate New York. While on these tours, Cole assembled a quite respectable collection of rocks and mineral specimens.[12]

Some of this devotion to the scientific study of nature rubbed off on the generation of artists Cole inspired. Cole gradually replaced formulaic conventions, techniques, and tricks that European artists had developed and perfected over preceding generations, developing instead what one art historian has called "a fresh response to the actuality of nature."[13] Though other art historians disagree, noting that Cole's practice was hardly a rejection of European conventions in favor of authentic experience of nature,[14] Cole did demonstrate to his fellow Hudson River colleagues how profitably one might spend days or even weeks in the field, carefully producing myriad sketches of fallen tree limbs and studies of curious rock formations. These studies could then be saved and used as reference materials in the studio, where their subjects were transplanted as the essential natural elements needed to adorn and give a unique vitality to the foregrounds of panoramic landscape compositions.

REPRESENTATIONS OF THE DELUGE

Amos Eaton's efforts to develop a comprehensive diluvial theory in the 1820s fit neatly into the global geological community's research program at the time. Art historians have long recognized that Cole knew about, studied, and conversed with geologists who were members of a generation that generally "placed their science in the service of religion."[15] Despite the potentially inflammatory impact of some early nineteenth-century geological findings (particularly those that challenged long-cherished religious doctrines regarding the date of the Earth's creation), art history scholarship suggests that such controversies tended to stoke rather than quash geology's popular appeal. Setting such preliminary conclusions aside, the subjects of contemporary geological controversy clearly had an immediate effect on landscape art, greater even than its effect on New York literature.

Artists on both sides of the ocean naturally took a special interest in the implications of new geological discoveries, particularly as they concerned the meaning of biblical scenes describing Earth-shaping events. By the late 1820s, the Deluge was revitalized as an especially attractive and significant subject for artistic reexamination.[16] One such portrayal of the biblical Flood, *The Deluge* (1826), by the widely admired English painter John Martin, triggered a flurry of public discourse and artistic response. In early 1829, a New York literary weekly reprinted a review that had been published the preceding October in the *London Monthly Magazine*:

> No artist of the present day can impart to his cause the impression of sublimity in natural objects that Martin does. If he paints a cavern, it is seen drawn out and expanded to the fullest size; yet its long-drawn, remotest recesses are brought before the eye with astonishing verisimilitude. Alpine precipices, and oceanic expanse, appear in their full dimensions on his canvass. Martin is the only artist we have ever seen whose pictures are stamped at first sight with the earthly sublimity of Miltonic poetry—"In the present work, he has summoned every aid to produce this great effect. The sun, moon, and a comet, are seen in

Figure 6 John Martin, *The Deluge*, 1828. Martin's painting was the provocation for Thomas Cole to attempt a more geologically informed rendering of the Deluge. It now belongs to the Georgia Museum of Art at the University of Georgia (University Purchase). Image taken from William H. Truettner and Alan Wallach, eds., *Thomas Cole: Landscape Into History* (New Haven: Yale University Press, 1994), 81.

conjunction. Noah's ark is placed on a lofty eminence which the waters have not yet attained; but its elevation bespeaks what the rising of the waters will be: this is a happy thought. The mingling of many conjunctive objects, some of them peculiarly terrible, to heighten the impression of the whole scene, displays powerful invention and a great judgment.[17]

According to American art historian Ellwood Parry, Cole took this laudatory statement personally; it was as if the writer had thrown down the gauntlet, challenging Cole to do Martin one better. The contents of Cole's sketchbook over the subsequent weeks and months attest to a new, vigorous obsession with producing his own depiction of the aftermath of the Deluge, one that would deviate from Martin's morbid fascination for momentary turbulence by focusing instead on the ultimate geological impact and significance of the Flood.

While Martin composed his imaginative rendering of the Deluge, geologists on both sides of the Atlantic were actively seeking to develop empirically detailed and theoretically sound analyses of that biblical event. Amos Eaton was a leading participant in this international quest to explain how and why New York State was covered by the various types and patterns of diluvial debris that he and so other many geologists had observed. In the brief interval between the publication of William Buckland's influential *Reliquiae Diluvianae, or Observations on the Organic Remains attesting the Action of a Universal Deluge* (1823) and the proliferation of various ice transport ideas such as those that Louis Agassiz and Charles Lyell would separately champion a decade later, it was especially compelling for a prominent American geologist like Eaton to try to confirm and extend Buckland's diluvial interpretation. To locate and scientifically document an antediluvian (or, more probably, an immediately postdiluvian) fossil human would represent a universally significant contribution to the discipline. Finding this desirable scientific object would help to confirm or deny the relevance of scriptural history to the geological record and at the same time (depending on where such a human's remains might crop up) would seriously complicate contending views regarding the standard accounts of human migration and inhabitation of the various parts of the globe.

Thomas Cole certainly knew about the debates that roiled geological discourse during the 1820s, if only through his casual conversations with natural historians at the Bread and Cheese club. His *The Subsiding of the Waters of the Deluge* (1829) unmistakably evokes Eaton's complicated diluvial hypothesis. Recalling for a moment how Eaton distinguished between "chaotic" and "mantling" stages of debris deposition in order to account for how a single inundation could possibly have wrought such a multitude of effects upon the land, Cole's painting far surpasses merely embracing either a watery Neptunism or a naive catastrophism.[18] Instead, he accomplished a distillation of the era's most sophisticated diluvial geological thinking. When facing the painting (Figure 7), the viewer seems to be situated within a cavern looking outward upon the world as it would have appeared during Eaton's second "quiescent" stage of debris deposition. The rocks have already been scraped clean of all soil and have been left in their jumbled state of disarray, with sopping bits of vegetable matter strewn about. In the foreground, Cole reveals the key scientific desideratum: that lone human skull. After its exhibition in 1831, one reviewer for the *New-York Mirror* remarked: "The grandeur of the subject strikes the mind with emotion, and the conception of the artist is highly poetic. The vast flood, gradually subsiding, leaves the peaked mountaintops visible; and the drenched world, as it again meets the light, has an air of deep solemnity and solitude extremely impressive. The effect is increased by the skull in the foreground. Fancy pictures the wretched relics, thus scattered beneath the waters."[19]

The perspective of peering out at the world from a cavern, which Cole adopted for this painting, would develop by necessity into its own tiny subgenre as painters flocked to the most geologically interesting natural history phenomena to be found in the United States. When the French-born and educated Marie-François Régis Gignoux traveled to document Kentucky's awesome Mammoth Cave in 1843, he composed a remarkable image that integrates accurate geological details

Figure 7 Thomas Cole, *The Subsiding of the Waters of the Deluge*, 1829. This painting now belongs to the Smithsonian American Art Museum, Gift of Mrs. Katie
Dean in memory of Minnibel S. and James Wallace Dean and museum purchase
through the Smithsonian Institution Collections Acquisition Program. Image taken
from William H. Truettner and Alan Wallach, eds., *Thomas Cole: Landscape Into History*
(New Haven: Yale University Press, 1994), 80.

with an uplifting gaze toward heavenly light. Placing the viewer in a deep dark
corner with a roaring fire in the foreground to light the cave's interior, Gignoux's
painting draws the eye through the contours of an almost throat-like cavern whose
limestone and sandstone surfaces reflect both the cumulative erosive power of
slowly dripping water and the glorious beams of daylight pouring in from the
cave's entrance at the very top of the canvas.[20]

ICONS OF DEEP TIME

Like Irving and Cooper had each done before him, Cole decided in 1829 to visit
Europe for a year or two. Bryant, who would make a similar journey of his own
in the mid-1830s, now published a poem intended to remind Cole of the superior natural features he was leaving behind. Though Europe's landscape might
betoken a glorious history, Bryant's sonnet boasts of America's "Lone lakes—
savannahs where the bison roves—/Rocks rich with summer garlands—solemn

streams— /Skies, where the desert eagle wheels and screams— /Spring bloom and autumn blaze of boundless groves."[21] Spending most of his time abroad in England and Italy, Cole went in search of refinement, professional connections, and a different kind of perspective so that he might return better equipped to capture the beauty and meaning of New York's special places.

Existing evidence also suggests that this period was when Cole became more serious about augmenting his practical field studies of nature by collecting and reading relevant geology books and articles. Historians of American landscape art make much of the fact that he owned a first edition of *Outlines of Geology* (1834) by John Lee Comstock, a prolific scientific textbook writer who was notable for promoting the controversial view that the "days" of Genesis may best be understood to represent indefinite periods of (geologic) time.[22] Interestingly, Cole was in England when he produced his first finished attempt to capture the power of Niagara Falls. His *Distant View of Niagara Falls* (1830) looks upstream at the cataract from the New York State side of the river, showing bright fall foliage softly surrounding the plunging river beneath dark skies and a pair of American Indian figures (one standing and the other crouching) upon an exposed promontory in the foreground. From the vantage point chosen, there is really no geological definition whatsoever displayed in the painting of New York's most well-known natural wonder.[23]

Cole's failure on this score was by no means atypical. Evidently, the play of water and mist was so overwhelming to the senses that none of his predecessors had yet paid significant attention to delineating the strata abutting the falls or along Niagara's rocky gorge. Louisa Davis Minot's *Niagara Falls* (1818), for example, presents a sidelong view of the cascade as seen from the American shore just beneath the waterfall. Though gently dipping slabs of rock and a handful of rounded boulders protrude across the foreground of the painting, Minot's tree-covered cliffs mask any suggestion of the erosive power of the turbulent waters.[24] Similarly, none of Alvan Fisher's spectacular trio of paintings, *The Great Horseshoe Falls* (1820), *A General View of the Falls of Niagara* (1820), and *Niagara: The American Falls* (1821), provide any useful information to the geological observer. The first displays a lovely rainbow arcing out of the gorge, as seen from the Canadian shore atop the Queenston bluffs.[25] The second, like Cole's a decade later, takes in the scene from such a distance downstream that no rock surface can be distinguished at all; from this perspective the cliffs (an unbroken wash of tan paint) might either be smooth chalk or sand.[26] The third of Fisher's attempts approached the brink but still offers only a uniformly blurry set of cliffs over which the river tumbles, sending up plumes of mist. The foreground of this final painting at least indicates how rocky the Canadian shore was below the falls, but its assembly of round boulders and nondescript stone surfaces was put there apparently only to provide a compositional substrate for Fisher's patterns of light and darkness.[27]

Through the 1830s, Cole increasingly incorporated particular natural phenomena into his allegorical paintings, and he tended to favor objects that constituted particularly provocative evidence in contemporary geological debates. Most notably, Cole's paintings featured those erratic boulders ("rocking stones") that had become so starkly visible in the deforested landscape of the rapidly industrializing states in New England and the middle Atlantic region. During this time Cole began composing his monumental series depicting Bryant's cyclical view of history, *The Course of Empire* (1834–1836). As one especially astute art historian has pointed out, a single immutable rocking stone draws the eye to the same focal point in all five paintings: "[Cole] actually employed a huge boulder, or rocking stone, or glacial erratic, precariously perched at the top of a great cliff (whose steeply inclined sedimentary strata are clearly indicated) as a mariner's landmark in each picture." Furthermore, since the finished cycle contains no riven mountains, nor any encroaching seas, in its fifth and final scene *Desolation* (1836), "the artist had decided to eliminate any sense that land forms can change as quickly as empires rise and fall."[28] This interpretation not only raises the possibility that Cole was a keen observer of geological controversies but also suggests that by the mid-1830s he (like Cooper) had become an early convert to Charles Lyell's uniformitarian ideas.

Cole also began to articulate an ideology and methodology for American artists that expressed the attitude Irving, Bryant, and Cooper had expressed as writers and paralleled the views Amos Eaton had been promoting for American geology throughout the preceding decade. Echoing his friend Bryant's sonnet, Cole exhorted his adoptive countrymen in his famous (1836) "Essay on American Scenery" not to be intimidated by Europe's "great theater of human events." Instead, they should go forth with confidence in celebrating the fact "that nature has shed over this land beauty and magnificence . . . [of a kind] unknown to Europe."[29] Meanwhile, Bryant had taken his family on a tour to Europe in beginning in the summer of 1834, and this trip gave him the opportunity to confirm with his own eyes what he and Cole had said about the comparative virtues of wild versus "civilized" landscapes. Upon seeing a painting by Cole of the Arno River, on exhibit in its native city of Florence, Italy, in October 1834, Bryant was characteristically moved to grieve for the impact of human artifice upon the land.

To him who views the vale of the Arno 'from the top of Fiesole' or any of the neighboring heights, grand as he will allow the circle of the mountains to be, it will appear a vast dusty gulf, planted with ugly rows of the low pallid and thin-leaved olive and of the still more dwarfish and closely planted maples on which the vines are trained. The simplicity of natural scenery, so far as can be done, is destroyed; there is no noble sweep of forest, no broad expanse of meadow or pasture ground, no ancient and towering trees clustering with grateful shade round the country seats, no rows of natural shrubbery following the courses of the rivers through the vallies. The streams, which are often but mere gravelly beds of torrents, dry during the summer, are kept in straight channels by means of stone walls and embankments; the slopes are broken up and disfigured by constant pruning and lopping, until somewhat more than midway up the Ap-penines, when the limit of cultivation is reached, and thence to the summits is a barren steep of rock without soil or herbage. The grander features of the land-scape, however, are beyond the power of man to injure—the towering moun-tain summits, the bare walls and peaks of rock piercing the sky, which with the deep irregular vallies, betoken more than any thing I have seen in America an upheaving and ingulfing of the original crust of the world.[30]

It is especially interesting to observe here how Bryant's Romantic geological sensibilities were revived by Italy's rugged mountains, even though that kind of dramatic language was more and more often derided as "catastrophist" by those among Bryant's intellectual circles who had come to embrace Lyell's views of gradual geological change. After he secured a lucrative commission for yet an-other epic series of massive paintings (*The Voyage of Life*), Cole returned to Eu-rope in the summer of 1841 in order to refresh his examination of Old World scen-ery. Many of the works he executed during his second European sojourn clearly extended and reinforced his appreciation for Lyell's uniformitarian approach.

Cole's *The Vale and Temple of Segesta, Sicily* (1844), for example, is more a tour de force in geological denudation than it is a paean to Roman ruins.[31] The temple occupies as much canvas space as his typically Lilliputian lone human figure. Light and shadow play across the entire vale, highlighting the stratigraphic details of at least four major rocky outcrops, boldly standing out against the back-drop of scantily vegetated stone surfaces that occupy the rest of the frame. These outcrops, which resemble knobby lumps enveloped by deep-cut ravines, vaguely hint at the volcanic origins of a bedrock surface whose coverings of soil and softer (sedimentary?) rocks have all presumably been worn away by the gradual but in-cessant action of water. Cole sketched detailed, on-site impressions of the Sicilian countryside in order to produce the finished work later in his Catskill studio.

Returning to New York in the summer of 1842, Cole resumed painting Catskill scenery with even greater attention to geological detail. Having seen Europe's most famous mountain ranges, he confided to the American consul at Rome: "Must I tell you that neither the Alps nor the Apennines, no, nor even Ætna himself, have dimmed, in my eyes, the beauty of our own Catskills? It seems to me that I look on American scenery, if it were possible, with increased pleasure. It has its own peculiar charm—a something not found elsewhere."[32] The affection comes across in his paintings. In dramatic contrast to the bleakly denuded Italian vale, his *Catskill Mountain House, The Four Elements* (1843–1844) and *Catskill Creek, New York* (1845) feature brilliant fall foliage and vibrant skies.[33] The boulders featured in the latter painting remind us of Cole's close study of erratics. Though he had been composing Catskill scenes for more than two decades, Cole remained dedicated to characterizing the topography, the stratigraphic patterns, and the rock compositions of rocky outcrops with care and apparent precision, enabling the viewer to discern whether boulders were or were not eroded exposures of local native rock. Of course, by this time geological theorists were vying to explain how the action of continent-sized glaciers (newfangled catastrophists like Agassiz) or floating icebergs (uniformitarians like Lyell), rather than inundations of extraordinary flood waters (old-fashioned diluvialists like Buckland and Eaton), might best account for the patterns of bedrock erosion and the transportation and deposition of massive alien rocks.

Cole continued to make research excursions to observe and paint significant natural history phenomena, such as coastal Maine's rugged Mount Desert Island, New Hampshire's White Mountains, and the waterfalls of the Genesee and Niagara rivers in western New York. Prior to his sudden (and quite unexpected) death in February 1848, Cole imparted his views and techniques to others. Besides writing his "Essay on American Scenery" and a few other contributions to the New York literary scene, Cole agreed to supervise the training of the next generation of American landscape artists. Frederic Edwin Church, an eighteen-year-old Hartford boy who was recommended by Cole's patron and hiking companion Daniel Wadsworth, became Cole's first student in 1844. Church apprenticed at the Catskill home of his teacher until 1846.[34] After his death, Cole's habit of direct study from nature did not wither as a quaint hallmark of the Hudson River School's founder; rather, it inspired colleagues like Durand, who embraced the practice to achieve even finer attentiveness to geological detail, and students like Church, who carried the scientific documentary mission along with him to the farthest reaches of the western hemisphere.[35]

CHAPTER NINE

Rocks, Reverence, and Religion

The champion of Geology[,] a sickly consumptive looking little
fellow, insisted upon the Science of Geology corroborating & proving all the statements and facts set forth in the Bible in relation to
the creation of the world &c and didn't seem satisfied until we all
admitted the truth of this assertion. I saw there was fire in the little
fellow so I denied it.
                    —D. C. Smith (one of Amos Eaton's students), 1835

The overlap between geology and religion constituted an important trading zone
for ideas in early American culture.[1] New discoveries in the earth sciences and
interpretations of the Bible not only stimulated public interest in the history of
the Earth but also catered to a growing multiform discourse on the meaning of the
American landscape. Scientific and religious belief systems sometimes clashed
over specific claims, though far more often these two competing forms of cultural authority were understood to complement each other as forms of knowledge
about nature. The boundary between professional and amateur scientist had not
yet been erected; mechanisms that would later partially insulate technical from
lay discourse (such as publication in peer-reviewed professional journals) had yet
to be developed.

Benjamin Silliman, the editor of the country's preeminent periodical containing technical reports in geology, for example, was himself a compromised gate-
keeper. In his editorial supplement to an 1833 revision of British agriculturalist
Robert Bakewell's widely respected 1813 text *An Introduction to Geology*, Silliman
blunted the author's apparently atheistic commitment to an Old Earth by cajol-
ing his American readers with these remarks: "In a country like this, where the
moral feeling of the people is identified with reverence for the scriptures, the
questions are often agitated:—When did the great series of geological events hap-

pen? If the six days of creation were insufficient in time, and the events cannot all be referred to a deluge, to what period and to what state of things shall we assign them? This is a fair topic of enquiry, and demands a satisfactory answer." After some further temporizing, Silliman reengaged with the hard nub of his task by taking a practical, almost sociological, slant: "The subject of geology is possessed of such high interest, that it will not be permitted to slumber; it will proceed with increasing energy and success; a great number of powerful minds and immense research are now employed upon it, and many collateral branches of science are made tributary to its progress." With diplomatic, if not entirely candid, aplomb, Silliman continued: "Its [geology's] conclusions have been supposed to jar with scriptural history: this is contemplated with alarm by some, and with satisfaction by a few; but there is no cause for either state of feeling: the supposed disagreement is not real; it is only apparent." Silliman concludes, with a kind of bravado that dismisses the key point of contention without really explaining how he can accept its implications and remain such a faithful Christian: "It is founded upon the popular mistake, that, excepting the action of a deluge and of ordinary causes still in operation, this world was formed as we now see it, and that all its immense and various deposits were made in a very short period of time. Both of these are fundamental errors, which have misled both the learned and the unlearned, and are still extensively prevalent."[2]

Silliman's mentor, Rev. Timothy Dwight, had foreseen the challenge that geology might pose to the tenets of established Christianity. Though he was supportive of the study of nature, and shared (and perhaps helped to nurture) Silliman's expectation that knowledge revealed through both geological investigations and contemplation of the scriptures would ultimately prove to be harmonious, Dwight in his own time had placed the burden of valid knowledge outcomes firmly on the scientist's intentions. One historian of American religion frames Dwight's caveat succinctly: "Scientific research was better carried out . . . with an attitude of trust in God's word and humility before his creation."[3] Thus, over the course of just a single generation, the balance of authority shifted rapidly from adherence to a religious orthodoxy toward acceptance of a scientific one, at least among those well-intentioned and earnestly devout practitioners of American natural history.

GREAT AWAKENINGS

Historians have long wondered about the conditions that fostered the birth of utopian cults (such as Mother Ann Lee's Shakers) and the spread of intense religious revivals (echoes of the first Great Awakening) near Albany from the late eighteenth century until the Civil War.[4] During and after the construction of the Erie Canal, New York State spawned a multitude of millenarian movements (most notably the Mormons and the proto–Seventh Day Adventist Millerites) alongside the backdrop of the Second Great Awakening's fervor (led by the charismatic Protestant revivalist preacher Charles Grandison Finney). Nobody who lived in New York State through these decades could have escaped encounters with the devotees of such strongly (often suddenly) held religious beliefs. To better understand the religious resonance of Amos Eaton's diluvial theories, and to make sense of the experiences that his students underwent as they ventured forth into the world to examine the land using the skills and intellectual preparation Eaton had provided, it makes sense to revisit the allure of ideas that were directly traceable to early eighteenth-century, Jonathan Edwards–style, New England Calvinism and that provoked such volatile ripple effects and reactions in New York throughout Eaton's lifetime.

Political scientist Michael Barkun attributes the social impact of renewed religious zeal in part to the success of the American colonists' efforts to wage a revolutionary war of independence. In the aftermath of that great military struggle, with the rise of indigenous political partisanship and a fracturing of the new country's sense of shared ideals, some of those who had understood the rebellion as a crusade for social change became disillusioned. While continued political struggle appealed to those who chose to carry the banners of Hamiltonian federalism or Jeffersonian democracy, others insisted that the struggle for salvation was necessarily a matter of cultivating individual virtue from within rather than constituting a collective social behavior from without. These contradictory impulses, according to Barkun, produced conditions convivial to a renewal of religious fervor over a prolonged period of time: "The post 1790-revivals were neither brief nor as spatially confined as their colonial predecessors. Dating the 'Second Great Awakening' permits simply a specification of its outermost boundaries and the years of maximum activity. On this basis it began about 1795 . . . and lasted until sometime between 1835 and 1860, depending upon the area of the country."[5]

Amos Eaton's generation was born during the war and matured amid the first rumblings of this resurgent religiosity. The oldest of Eaton's surviving private journals describes in introspective detail the climate of devotion to which he was exposed as an impressionable youth in the tiny Columbia County hamlet of New Concord, New York. At his father's insistence, at the age of eleven Amos spent the year of 1787 witnessing and pretending to partake of the revivals. By the following spring, the boy had succumbed to persuasion: "I was really and sincerely engaged in Christianity. Never in my life had I felt so happy." His personal religious faith would wax and wane over the subsequent years, but his fundamental devotion to God was firmly rooted in this formative coming-of-age experience. At seventeen, he wrote that he considered the purpose of his education to be preparation to preach the gospel.[6] His early sense of a calling to the ministry was soon derailed into the futile pursuit of a career in law, followed by imprisonment, penance, and rebirth as a natural scientist and teacher. In this roundabout manner, he found a way four decades later to make good on his adolescent pledge.

As the senior professor at the Rensselaer Institute from the mid-1820s through the 1830s, Eaton was responsible not only for science classes but also religious instruction. Each week on the Sabbath, he took the lectern to teach "sacred history" from his vantage point as a geologist. One student's notebook from 1836 records the contents of Eaton's successive weekly sermons. One Sunday, the professor detailed the first two "Periods of Time" on the Earth. Namely, these were the "16.56 *ages*" lasting from the Creation to the Deluge and the 417 years that elapsed between the Flood and the "Calling of Abraham." Eaton explained that different animals roamed the world before the Deluge, citing as evidence the diluvial deposits containing fossil remains far different from the animal species that inhabit the modern world. To account for the mystery of why certain species of fish were apparently extinguished by the Flood, he speculated that many marine creatures died from a sudden temperature change in the oceans. According to his diluvial theory, he supposed that the initial "violent phase" of the Deluge would have intermixed colder deep sea waters with shallow coastal waters, creating heat conditions severely inhospitable to the denizens of every clime.[7] In his remarks the following Sunday, Eaton proceeded to identify specific locations around Troy that students could recognize as being either diluvial or antediluvial in aspect. Admitting that no products of human artifice had yet been found among what he called New York's diluvial debris, he held out hope for some such evidence to materialize because "the inhabitants of the earth before the deluge led a pastoral life."[8]

If Eaton's geological ruminations displayed their most explicitly theological cast on Sundays only, those of his leading contemporary colleagues were even more consistent in their Christian allusions. In Connecticut, Silliman publicly defended the scripture's relevance to geology, and the first state geologist of Massachusetts was a man of the cloth, Rev. Edward Hitchcock. After Hitchcock was inspired to study geology by one of Eaton's itinerant lectures in Northampton in 1817, he followed Eaton's example and studied at Yale with Silliman. Hitchcock had more than caught up to Eaton professionally by the time the Rensselaer School was founded in Troy. Hired in 1825 to run the state geological survey as well as to serve as professor of geology and chemistry at Amherst College, Hitch-cock would complete the first comprehensive geological survey report of any state or comparable political division in the world by 1832.[9]

Hitchcock's first published paper on the geology of the Connecticut River valley in Massachusetts showed that he was willing to take a public stand far more attuned to the religious dynamic of natural history than Eaton was. Ironi-cally, like Eaton's notice on the "singular" deposit of gravel recorded that same spring, Hitchcock's paper (discussed in chapter 1) proposed that each winter's accumulation of ice-eroded rocks might constitute a "cosmogonical chronom-eter." Like others who studied the recession of Niagara Falls, Hitchcock sought to deduce the age of the Earth from those heaps of shattered rocks. Proceeding to attack early suggestions that the Connecticut valley may have been filled by vast lakes in former ages (ideas first proposed by observers like DeWitt Clinton but later revived by devotees of the glacial hypothesis), Hitchcock advanced his own catastrophist theory by developing an argument that ingeniously married his fidelity to scriptural truth to very careful empirical scientific reasoning. "For every geologist knows that all this [detritus of older rock formations] must be referred to a period anterior to that, in which the last great diluvian catastrophe happened to the globe and left our continents in their present form. Nor is the mere occur-rence of masses of stone, evidently rounded by the attrition of running water, any evidence in favor of this [former lakes] hypothesis; for we must look to the cause of this also, as far back at least as the Noachic deluge.—No current of water with which we are acquainted is sufficient to transport such masses of rock into the situations in which we find them."[10]

Systematic diluvialism would enjoy its apogee in English and American geo-logical circles in the mid-1820s. Eaton and Hitchcock led the way for Americans interested in exploring the scientific veracity of the biblical deluge. Their intense interest and extraordinary patience in tracing out what they supposed must have been the Flood's elaborate manifestations was shared by some of Britain's lead-ing geological writers. John Phillips, who incidentally was married to the sister of the famous English canal engineer and stratigrapher William Smith, asserted with confidence in 1829: "Of many important facts which come under the con-sideration of geologists, the 'Deluge' is, perhaps, the most remarkable; and it is established by such clear and positive arguments, that if any one point of natural history may be considered as proved, the deluge must be admitted to happened, because it has left full evidence in plain and characteristic effects *upon the sur-face of the earth*." Phillips also echoed Hitchcock's earlier assertion that currently observable processes (the key to the uniformitarian principle of "actualism") were inadequate to explain all of the diluvial debris: "It is impossible to account for the vast heaps of this gravel by supposing that it might be laid in its present situation by any streams such as now water the earth. For it occurs abundantly in places where streams do not run, where, indeed, they never did run; neither is it confined to such narrow paths as serve for the passage of rivers, nor is it laid in such forms, but it is casually and unequally spread over all the face of the country." Awed with the scale of erratic boulders, Phillips continued: "The blocks of stone which have been rolled from their native sites, are, in some case[s], of so vast a magnitude, and have been so strangely carried, that in vain do we think to assign any other cause for the phenomena, than a great body of water moving upon the earth." Finally, however, Phillips (like William Buckland) was more guarded than his American colleagues when it came to associating any specific diluvial phenomena with Noah's Flood: "Those geologists have been ill-advised, who, in the present state of the science, affect to form a chronology of nature for comparison with the records of history."[11]

Throughout the 1820s Eaton, Hitchcock, and Silliman all championed the primacy of water as the geological agent most responsible for shaping and reshaping the surfaces of New York and New England. Each man subscribed to diluvial explanations for the salient phenomena they encountered in their research travels. As Rodney Lee Stiling notes, when Silliman attached his "Outline of the Course of Geological Lectures, Given at Yale College" to his 1829 edition of Bakewell, the section concerning the Genesis flood deviated most sharply from the original author's text.[12] Silliman believed that the American reading public was particularly concerned about the religious implications of geological theories. Even so "practical" a geologist as Eaton habitually subjected his thoughts to a Christian religious filter. For example, when privately contemplating geological evidence of catastrophic extinction events, Eaton draws upon biblical imagery in composing an entry for his geological journal during the summer of 1829. Incidentally, this passage presages the climate shock reasoning (which Eaton would later share with his students) to account for why certain marine species did not survive Noah's Flood:

It is a remarkable fact, that most antidiluvial [*sic*] animals were of the strong carnivorous kinds, of the most ferocious character. It would seem that the spirit of Cain not only infused itself into his successors of the human race, but into the beasts of the forest, and into the whole amphibious and aquatic tribes. The shark and the lizard and all were overthrown by the deluge and engulfed in the same common tomb with man. [French Geological Society founder Louis]

Cordier should assign for the cause of this ferocity, the high temperature of the earth in those days; and should show us how the cold washing of the Earth, by diminishing its temperature, moderated the fury of its inhabitants.[13]

Neither Eaton nor Hitchcock were totally theory driven or closed-minded about their scientific views. Their ideas grew and changed significantly over time, and stimulating discoveries provoked them to formulate innovative theories. Eaton, for example, modified his early strong predisposition in favor of watery explanations after closely examining the Palisades along the western shore of the Hudson across from New York City. His field notes record the equivalent of a "conversion moment," when the persuasiveness of a Huttonian-style, heat-driven metamorphism became undeniable and added a new and useful contributory mechanism to the old Wernerian catastrophist's reasoning.

On examining the Palisades, it appears probable, that they were made by immense subterranean heat, which broke up and melted the primitive rocks of the Highlands, causing the chasm through which the Hudson flows. That the melted lava ran down in a southerly and southwesterly direction forming the basaltic range to near N. York and then winding southwesterly across N. Jersey &c. while running over the variegated sandstone, sometimes partly fusing, and intermixing with, it. Probably the ridge was formed under water, which became consolidated in columnar forms; while some was left in a semi-indurated state, and being washed away left a channel on its eastern side, along which the Hudson began to flow as soon as the oceanic waters retired.

Suddenly captivated by the power and elegance of this different way of imagining the Hudson River's formation, Eaton immediately tried to apply the same formula to other puzzling phenomena he had observed: "The same hypothesis will apply to the range of basalt hills from the north line of Massachusetts along the Connecticut River to Middletown in Connecticut, and thence to New Haven. For the chasm, now occupied by Connecticut River, would have afforded ample materials for the necessary quantity of volcanic lava for all of those hills, together with all that may be supposed to have been washed away while in a soft state leaving the hills in their present insulated situations." Jumping finally to the broadest possible generalization, Eaton closed his journal entry with a characteristically provocative query: "Cannot all basaltic series of hills, mountains, and mural [escarpment] ranges, be explained by refference [sic] to chasms, whence their fused material may have originated?"[14]

In much the same manner, Hitchcock was perplexed when he tried to analyze dinosaur footprints that had been found in the petrified mud of the Connecticut River valley. By the mid-1830s, he promoted a hypothesis that attributed these fossils to gigantic, flightless, prehistoric birds.[15] Although Hitchcock lived long enough to become a steadfast opponent of Charles Darwin's views on evolution, he did move gradually and graciously to embrace ice as the most likely explanation for how erratic boulders had been transported across the North American continent.[16]

Despite all the startling findings about the Earth's deep past and the many substantial modifications their geological ideas would be forced to undergo, neither Eaton, Hitchcock, nor Silliman were ever fundamentally shaken in their basic religious convictions. Throughout his long career Hitchcock especially managed to attract and retain the respect of fellow scientists and of clerics, the vast majority of whom remained sympathetic to, if not utterly convinced of, ultimate harmony between religion and science. With a sophistication and forthrightness scarcely to be found among Christian opponents of Darwinian evolution a century later,[17] Hitchcock arrived at the empirical scientific conclusion that Noah's Flood, though unquestionably (for him) a real historical event, may in fact have left behind no absolutely unique geological consequences. Hitchcock's 1854 edition of his great tome *The Religion of Geology* frankly concedes this point: "From the facts that have been detailed, it appears that on no subject of science connected with religion have men been more positive and dogmatical than in respect to Noah's deluge; and that on no subject has there been a greater change of opinion. From a belief in the complete destruction and dissolution of the globe by that event, those best qualified to judge now doubt whether it be possible to identify one mark of that event in nature."[18]

PUBLIC ENCOUNTERS WITH PROVOCATIVE
GEOLOGICAL CLAIMS

Whenever Eaton's students were away from school, they were encouraged to write letters to him describing the places they visited and the natural history specimens they gathered. Eaton had instituted this practice from the very first class he taught at Williams College in 1817, and it had been one of the key tools that had enabled him to draw together a comprehensive picture of the botany, mineralogy, and geology of considerably more territory than even he had been able to cover in two decades of exhaustive research travels by foot, boat, or carriage. On one such occasion, one of Eaton's students penned a particularly illuminating account of his experiences aboard a public coach, in which he provoked and then witnessed the unfolding of a dispute over the competing truth claims of geology and "anti-geology." (The opening of the account provides the epigraph for this chapter.) A detailed analysis of the episode can perhaps help the twenty-first century reader to recapture the visceral intensity with which citizens in the broader culture were grappling with the religious implications of contemporary geological discoveries and theoretical pronouncements.

D. C. Smith spent much of the summer of 1835 touring the New England states and Quebec province. His tour commenced with a visit to the historic Charles Bulfinch–designed state house (where the Hartford Convention had met in 1814, a spot that commemorated Federalist demands that New England secede from the United States rather than continue to fight the locally unpopular war against Great Britain). Here Smith befriended a fellow tourist who was the son of bishop Thomas Church Brownell (probably his eldest son Thomas, who was nineteen at the time but would only survive another six years). The young Brownell offered to be Smith's guide in his travels upstream along the Connecticut River.

Eaton would certainly have been impressed by this stroke of good fortune. His student's traveling companion had a respectable upstate New York scientific pedigree. Bishop Brownell had been educated at Union College in Schenectady, graduating at the head of his class in 1804. Union College then hired its talented alumnus to stay on and teach Latin, Greek, belles lettres, and philosophy. When Union decided to break new ground and become the first local college to offer a course of study in the natural sciences in 1809, the senior Thomas Brownell was the faculty member appointed to serve as the new professor of chemistry and mineralogy. Like Silliman, Brownell was given one year's leave to acquire scientific training at European institutions. Due to the ongoing hostilities of the Napoleonic Wars, his intellectual sojourn was confined to the British Isles, where he studied with Humphrey Davy in London, examined the crystalline basalt of Ireland's Giant's Causeway, and visited laboratories and audited lectures by some of the leading Scottish Enlightenment scholars then working in Glasgow and Edinburgh.[19] Brownell might easily have gone on to eclipse Eaton as the Albany area's leading practitioner and teacher of the natural sciences had he not met and married Charlotte Dickinson of Lansingburgh, New York, in 1811.

An ardent Episcopalian, Charlotte stoked the fires of Thomas Brownell's brewing discontent with Calvinism, even in its "mitigated form."[20] Living in the midst of the Second Great Awakening, Brownell found himself persuaded to abandon his professorship in order to study theology. Ordained in 1816, he became rector at New York's Trinity Church in 1818. In 1819, the Connecticut diocese surprised Brownell by electing him to fill the vacancy created six years earlier by the demise of bishop Abraham Jarvis. Brownell relocated his young family to Hartford and looked for ways to combine his vocation as an educator with his new responsibili-ties as a church leader. He was frustrated when the General Theological Semi-nary flirted with relocating from New York to New Haven in 1820, a move that failed to become permanent. After this disappointment, Brownell took advantage of changes in the new Connecticut state constitution, which in 1818 had loosened the Congregational church's exclusive grip on political power, and secured a charter in 1823 to found a new sectarian institution of higher education (Wash-ington College, later renamed Trinity College). Brownell served as the college's first president from 1824 until 1831. In the meantime, without entirely abandoning his interest in natural history, the bishop led a missionary tour from Cincinnati to New Orleans in 1829, recording along the way observations of the geology of Kentucky, Alabama, Mississippi, and Louisiana in his journal.[21]

As Smith and the younger Brownell headed north along the banks of the Connecticut, they were getting along agreeably with their fellow coach travelers until a pair of passengers suddenly erupted into a desperate argument. Smith intervened to question the initial extreme claim offered by the "champion of Geology." The conflict soon spiraled past his polite objection and then careened beyond all bounds of civility: "The denial was scarcely out of my mouth before Anti-Geology finding that he was not alone in the opposition, fastened on to the little fellow in a regular way, they quoted Scripture & Geology. I occasionally brought to bear a shadow of common sense, this but seldom, however untill [sic] they had worked themselves into a flaming passion, to the no small gratification of the other passengers, and from the lie circumstantial they went to the lie direct when Anti-Geology finished the argument by knocking the little fellow down, and injured him severely."[22] Why should ordinary people resort to fisticuffs in a public space over an essentially theological disagreement? Neither participant was ques-tioning the truth of the Bible. None present took the position of "Anti-Religion" in this debate. They merely disputed whether the findings of geological science were (or were not) a reassuring or reliable confirmation of what the Bible asserts about the Earth's early history.

If the account written by Eaton's student had simply terminated here, one might write off the entire episode as just a curious eruption of tempers among individuals who may or may not have had any philosophical understanding of the scientific facts they were debating. The letter goes on, however, to describe how the fight was broken up. The two men were restrained by other passengers, and when the stage arrived at its next scheduled stop in Brattleboro, Vermont, the aggressor who threw the first punch was persuaded by the assembly to promise the "little fellow" five dollars rather than be ejected from the coach. The two men then rode on in gloomy silence. Smith and Brownell finally exited at Walpole, New Hampshire. Smith told the pugilistic Anti-Geology that he would have given a good deal more than five dollars to settle the affair. This fellow's retort demonstrated that he knew a thing or two about Hutton's volcanic theory of rock formation. According to Smith's paraphrase, "if Granite was thrown from below to the surface of the earth by convulsions of nature, he [Anti-Geology] thought that after the opening of the sixth seal [a prophecy from the Book of Revelation] the commerce of N[ew] Hampshire would be mostly destroyed as their staple commodity would be easily found most any where."[23]

To recapitulate, the argument on the passenger coach escalated when Eaton's student piped up with a modest objection to the strong claim advanced by the champion of Geology—that all theological truths might be provable on the basis of geological evidence. This position represented something of a rebellion against his teacher's instruction, for Eaton himself had been sincerely engaged in demonstrating the harmony of religion and science, even to the point of sometimes using geology to "corroborate . . . facts set forth in the Bible." Eaton, Hitchcock, and Silliman all considered the Bible more than just a concomitant source of information; the scriptures were also an occasionally necessary corrective to purely empirical forms of scientific reasoning. On the other hand, none of these professors would probably have chastised young Smith for taking such a challenging position in the debate. For starters, the courteously framed challenge need not have provoked a vicious bloodletting; had it been posed in Eaton's laboratory or abroad on a field trip, an edifying intellectual discussion would surely have ensued. Moreover, Smith was not mischievously playing the devil's advocate. His objection, rather, espoused the conservative attitude of early nineteenth-century scientific Baconianism, which insisted that inductive arguments cannot aspire to any explanatory power beyond the reach or quality of the body of data thus far gathered. As Smith explained to Eaton, he understood that "common sense" was an important basis for proceeding through the kind of philosophical minefield into which he and the other passengers aboard the coach had trespassed. This episode clearly illustrates how poorly the lines of demarcation were drawn between the proper realm of geological and theological inquiry in 1835. In the absence of any general consensus about such matters, a kind of heuristic social etiquette demanded that traffic through the "trading zone" ought to be guided by prudence and policed by a respectful disengagement from direct conflicts between faith and reason.

A tantalizing sense remains, hinting at the complexity of interactions between theoretical debates in geology and lay understandings of the landscape. People really did care about the answers to the questions geologists were asking. How did the rocks under all our feet come to be and come to rest where they are? By what processes, and within what kinds of time frames, had the landscape been formed? Clearly, the implications of these questions for doctrinaire religious beliefs were manifold. Nevertheless, the whole affair also shows that the popular image of warfare between science and religion, problematic as it is in depicting post-Darwinian controversies, is even more misleading with regard to these earlier tensions that geological discoveries about deep time began to pose for certain kinds of claims on behalf of biblical literalism.[24] Historian Walter Conser provides a sensitive reading of the shifting balance between scientific and religious authority in antebellum America, and he documents such a diversity of interactions that any simple reduction of the story to one of conflict or of harmony becomes quite untenable.[25] As an example of positive collaboration in scientific knowledge production, another historian who specializes in the various American diluvial theories asserts that "theologically orthodox . . . Christian geologists *led the way* in taking Americans to a local view of the Genesis flood."[26] In other words, geological speculations about a biblical Flood were far from being obstacles to understanding the natural environment, even though they were necessarily constrained by theology-laden assumptions. These early efforts would later suffer the scorn of subsequent generations of geologists, who imagined that their own theories were far less metaphysical than the embarrassing Bible-driven fantasies of their predecessors.

The utility of any theory, however, is measured neither by its immediate accuracy nor its longevity. Rather, a theory is good if it is *fruitful*, calling attention to significant natural phenomena and spawning new explanatory mechanisms and techniques that practitioners can use to modify and improve their overall understanding. This critical process of inquiry often requires modification or even rejection of the initial theory's assumptions and framework. By these criteria, the kinds of diluvialism that Eaton, Hitchcock and Silliman promoted comes out looking pretty reasonable—surely deserving of more credit than their students were subsequently able to appreciate after undergoing their own profound conversions to uniformitarian methods and glacial theories. For early nineteenth-century New Yorkers, the salient phenomena requiring explanation were those massive erratic boulders, the jumbled heaps of gravel and sand, and the parallel grooves etched in solid bedrock that suggested episodes of past violence in the landscape. The investigation and elaboration of those ingenious diluvial scenarios, which Eaton and others inventively imagined and refined throughout the 1820s, prompted a lot of extremely valuable surveying and geological mapping. Eventually, the data gathered through this arduous fieldwork would serve to document the paths of the ancient ice sheets that sculpted and redistributed soil and rocks across much of the North American continent.[27]

While early American geologists opened up new possibilities as a result of all this exciting and creative work, they also exposed divergent tendencies within the practice of science itself about how best to generate fruitful, testable theories and explanatory mechanisms. What was at stake was not simply the scale of prehistoric time nor the nature of the causes that have ever been involved in shaping the landscape, although these were indeed troublesome issues insofar as prevailing religious belief systems were concerned. Alongside these metaphysical and ontological matters, however, geologists in New York and New England in the 1820s and 1830s were putting their methodological and epistemological assumptions to the test. As the scope of geological time suddenly grew from thousands to hundreds of thousands of years, should the testimony of recorded history (such as the Hebrew scriptures, which though ancient and revered, were at best a few thousands of years old) continue to play such a pivotal role in the interpretation of empirical observations from the field? What kinds of ideas were going to prove robust and resilient in the face of new discoveries made as other localities of the North American continent began to enjoy the same degree of trained scientific investigation that the northeast was already experiencing? Finally, there was an important strategic political question that savvy scientists (like Amos Eaton) were usually careful not to ignore. What kinds of arguments were likely to promote rather than hinder public support for institutions that would ensure the establishment of professional scientific authority and success?

The transformation of geological discourse in the decades following Charles Lyell's promotion of the uniformitarian method can perhaps best be understood in terms of the considerations listed above. Lyell advocated both actualism and "gradualism" (a view that preferred explanations premised on slow accumulative processes rather than claims of sudden catastrophic geological change) as essential components of his system. These attitudes promised to bring uniformity and rational order (lawfulness) to the earth sciences. Given the preceding century's provocative breakthroughs in Enlightenment natural history, the unsettled boundaries of geological expertise vis-à-vis more traditional forms of authority, and the insecure theoretical knowledge base and fluctuating practices that allowed innovators like Eaton to thrive (even though he was effectively a defender of Christian orthodoxy in American geology), it makes sense that Lyell's ideological methodology had a precipitous impact on the entire field within a very short period of time.

For example, Louis Agassiz's ice age hypothesis was initially inadmissible for

uniformitarians, who considered it just another catastrophist appeal to unobserv-
able forces acting on an outrageous scale. The ingenious theory was eventually
rendered acceptable to uniformitarian geology as a result of a two-stage process
that involved meeting each of these criteria. According to actualism, only currently observable causes could be assigned any responsibility for past events that
might have shaped or reshaped the Earth's surface. Therefore, until the 1850s,
when explorers finally penetrated far enough inland on the island of Greenland
to observe the existence and monitor the dynamics of a massive ice sheet, flooding was the only natural agency people had witnessed whose power seemed even
remotely adequate to have transported rocks and debris in the sizes, quantities,
and distances required.

Given its strong preference for gradual change, uniformitarian analysis naturally tended to amplify the duration of terrestrial time. Even the diluvialists of the
1820s (like Buckland and Eaton) had moved inexorably toward the view that not
just a single flood, but many stages of flooding, would have been needed to account for all the different phenomena they regarded as diluvial. There was growing consensus that the Earth's past history was far longer than the few thousands of
years countable in terms of the lives of the biblical Adam and his finite generations
of descendants. Therefore, the tremendous scale and intensity of violence assigned
to past Earth-shaping events could subside into explanations wherein a stupendous
canyon could be carved over countless millennia by a tiny stream; geologists no
longer had to insist upon some "lake bursting through mountains" scenario or a
supernatural fit of temper to account for the dramatic sundering of a rocky chasm.
With respect to the ice age hypothesis, therefore, once the dynamics of Earth's past
climate variability had been linked to evidence not only of glacial scratching but
also some kind of continuity with equally dramatic evidence that tropical conditions had sometimes prevailed at the higher latitudes, the "normality" of an Earth
passing gradually through cycles of heating and cooling came to be accepted on a
geological scale of time that involved millions of years rather than thousands.

Naturally, for the old defenders of Christian dogma, the new methodology
caused concern on a variety of fronts. Eaton managed rather gracefully, however,
to accept bits and pieces of Lyell's terminology; of particular significance was the
Englishman's systematic use of the term *Tertiary* to describe the same variety of
clays and sands that lay above Eaton's "upper secondary" formations. By 1841,
Eaton had agreed to replace his own diluvial category with the word *Tertiary* in
all instances.[28] Hitchcock, for his part, was highly skeptical about Lyell's radical
insistence on actualism. Anxious about the Englishman's apparent lack of a religious perspective, Hitchcock held out for a long time against the idea that God

might not have invoked greater uses of natural forces in the past than anyone was able to observe in the present. Hitchcock wrote to Silliman in 1837, confiding his doubts while flattering his righteous friend: "If he [Lyell] be a decided believer in Christianity why . . . does he not say so? A single sentence would settle the point. — Did anyone ever read the geological writings of Prof. Silliman and doubt whether they were the friends of revelation?"[29]

On the other hand, the prevailing devoutness of American geologists during the first half of the nineteenth century allowed for a sometimes too tolerant attitude toward the injection of religious opinions into technical scientific matters. As Eaton was quick to realize, the construction of the Erie Canal provided an opportunity for others to do groundbreaking, internationally respectable geological research. The influx of so many people through the terrain he had analyzed would also expose his ideas and reputation to diverse challenges. By 1828, Eaton's patience had been overstretched by listening to too many fellow canal boat passengers as they presented their idiotic theories of geology to him uninvited. He went so far as to vent his irritation publicly, though he was at least prudent enough to bury these caustic remarks deeply in a technical report on geological nomenclature: "The absurdities of Ditton, Whiston, and Buffon [seventeenth- and eighteenth-century European "armchair" natural philosophers] are outdone by our travelling Werners and Huttons [enthusiastic Americans who think they have read up on modern geological theory]. Even grave looking clergymen and civilians, who cannot distinguish granite from puddingstone, often hold us by the button to tell us how a ridge was thrown up, or a rock thrown down, or how a breach was made through a mountain."[30] For Eaton, the skills of a practiced field geologist were far more relevant than adherence to any kind of religious dogma in rendering one eligible to engage in the imaginative tasks of explaining how those rocks had been moved.

A TIME FOR PASSING AWAY

For Amos Eaton, the magnificent mystery and the laborious difficulties of decoding the Earth's past all came to an end on 10 May 1842. He died just one week shy of his sixty-sixth birthday. A former student would include this elegaic stanza as a part of his poetic tribute to the fiftieth anniversary of the founding of the Rensselaer School: "Sleep noble Eaton / in thine honored rest, / No anxious cares to pain thy peaceful breast, / But grateful words in granite shall proclaim, / Our lasting reverence for thy worthy name."[31] At the same festive occasion, the prominent lawyer and politician Martin Townsend spoke at length about what Eaton and his fellow geologists had contributed to the world of human understanding.

Comparing Eaton's efforts to decode the Earth's history to Jean-François Champollion's famous work to unlock the secrets of Egyptian hieroglyphics, Townsend expounded in 1875 about New York's early geologists in a language and manner that blends the deep time imagery of William Cullen Bryant and James Fenimore Cooper with an unabashedly religious tone of benevolent purpose. "They have shown us that in our walk upon the earth we are treading amidst a vast extended grave yard, where the bodies of the animal dead have in successive generations been buried and stored away as in vast museums for millions of ages—yea, for millions of ages before the primeval man was fashioned by the hand of the Creator, and they have opened up to us the exhaustless storehouses in which the early animal dead have so long rested in fossil forms. . . . What tongue can adequately describe the world of wonder the naturalists have opened up to us?"[32]

Anyone with the barest acquaintance with the history of science might suppose that a necessary conflict between religious orthodoxy and scientific truth emerged abruptly and spontaneously from the publication of Charles Darwin's *The Origin of Species* in 1859. Readers who study this topic soon learn, however, that a broader cultural awareness of questions regarding Earth's chronology (deep time) predated Darwin's theory and that it was the geologists throughout the preceding century whose discoveries were most responsible for placing a gradually accumulating strain on the credibility of a literal interpretation of the Christian scriptures.

In early nineteenth-century America, the champion of geology was not necessarily an enemy of religion. Rev. Edward Hitchcock built his entire career on a foundation that presumed that an ultimate harmony must exist between scientific understanding and scriptural teachings. Many aspects of his knowledge of geology underwent acrobatic changes over the course of his lifetime; his flexibility can be seen in his serial endorsements, in turn, of Abraham Gottlob Werner's neptunism, a compromise view that incorporated James Hutton's vulcanism with respect to primitive rocks like granite and gneiss, William Buckland and Amos Eaton's visions of diluvialism, and elements of Charles Lyell's uniformitarian methodology and Louis Agassiz's ice age theories. Yet throughout all these modifications of scientific opinion, Hitchcock never surrendered his deep conviction that the process of scientific investigation would bring him closer to truth and to God. Eaton's religious devotion may never have equaled that of his friend Hitchcock and he was generally slower to adapt to new theoretical fads in geology (unless of course he was their author), but it is valid to say that Eaton spent the second half of his adult life investigating and teaching others about New York's geology without ever agonizing whether this subject might cause him to abandon or be isolated from his generation's almost universal faith in a Christian God.[33]

Echoes of New York's Embrace
of Geological Investigation

During his lifetime, Amos Eaton witnessed and contributed to a radical transformation of theory and practice in American geology. Contrary to what one might suspect from reading ordinary American history textbooks, the first four decades of the nineteenth century were an era in which scientific investigations were widely considered to be relevant to the broader needs and intellectual interests of the general public. Collectively, the political, literary, artistic, and religious discourse of these decades displays a wide variety of dramatic and colorful examples of lay awareness of contemporary geological ideas and hints at wide participation in debates that arose from those investigations.

This book has examined how changes in geological knowledge both derived from and enabled the rise of the Empire State. Between 1776 and 1842, New Yorkers worked hard to develop respectable representation among the new nation's cultural institutions and were leaders in American commerce and technology. New York's successful multivalent exploitation of its physical geographic advantages depended essentially on the now little considered fact that its early political leaders, including the Jeffersonian Republican DeWitt Clinton and the Federalist Stephen Van Rensselaer, were willing to work together to make serious investments in institutions and projects that would foster local talent in the study of natural history. The centerpiece of their collaboration was, of course, the construction of the monumental Erie Canal, which provided the easiest avenue for internal immigration from the Atlantic seaboard to the continental interior while effectively capturing the lion's share of Great Lakes commerce through New York State. In all the reams of writing one can find celebrating this remarkable technical feat, very little consideration has been paid to the significant ways in which local knowledge of New York's physical geography and geology made completion of this audacious project financially and politically feasible. When one considers the many frustrations and material setbacks experienced by canal promoters during those seven years of construction, one sees the political fortitude and technical ingenuity required to sustain multiple phases of canal excavation through a variety of challenging terrain. The project might easily have proven just as foolish and unprofitable as the grandiose canal projects subsequently launched in every state that tried to imitate New York's astounding initial success.

Even among scholarly historical studies of the early American republic, it is rare to see a critical analysis of the circumstances that elevated the pursuit of theory and practice in the natural sciences to a place of professionalism, high respect, and social authority.[1] Paradoxically, while it has become almost a cliché to attribute the emergence of a distinctly American national identity to the vastness and sublimity of American nature, the research literature that examines just how Americans used the land to formulate a distinct national character has developed as a haphazard patchwork of subspecialties. This book has drawn upon work by historians of science, technology, political science, literature, and art in order to better comprehend the activities, institutional innovations, individual contributions, and, most importantly, the intersecting relationships across disciplinary boundaries that were ultimately responsible for New York State's leadership in the development of new intellectual capacities and widespread cultural appreciation for the scientific study of nature.

Three familiar tropes of American intellectual history have informed this study: the struggle for independence from European political and cultural hegemony, the frontier reputation for placing practical considerations above theoretical ones, and the metaphor of warfare between science and religion. Though each of these tropes possesses at least a germ of enduring truth, in terms of my specific findings, all three historical generalizations require some fine-tuning and modification. The most fruitful and robust of the three is definitely the struggle for American independence from European political and cultural hegemony. A respected elder colleague of mine in the history of science once shared his opinion regarding how this trope applied to the special case of geology: "Every intellectual endeavor in America derived from and emulating European activities had to generate its own sense of independence, even as had our political system. In the realm of science, America was in a colonial state of dependency. In natural history and geology however, the subject matter was necessarily distinct from that in Europe and the sense of nationalism rather easy to develop."[2]

In the particular case of Amos Eaton's life and career, his persistent desire for international acclaim and collegial respect complicated the creative impulse to innovate. Eaton's theoretical innovations and his predilection for assigning new names for rock layers could provoke harsh ridicule, especially from someone like George William Featherstonhaugh, a foreign-born scientist who traveled in the same New York scientific institutional circles and who competed directly with Eaton for the favor of Van Rensselaer's generous patronage. To be treated like a member of an inferior culture was, for Eaton, politically and psychologically intolerable. On the other hand, the daunting practical challenge of trying to reconstruct the planet's past absolutely required an open attitude toward transnational cooperation. It also required the general sharing of knowledge and practices, agreements on analytic terminology, and concerted efforts to isolate and make sense of localized natural phenomena that could help to identify and illuminate the effects of geological processes that may have operated regionally or globally.

American ambivalence toward European scientific societies was manifested in contradictory behaviors. Institutions like London's Royal Society and the Académie des Sciences in Paris were historically associated with the monarchical form of government against which Americans had waged two wars already. Even so, being nominated to hold a corresponding membership in one of these societies was a highly coveted honor in New York, as exemplified by DeWitt Clinton's and David Hosack's celebrated elections to Britain's Royal Horticultural Society in 1817. Visiting European members of these same societies had to go to great lengths, however, to win the esteem of their North American colleagues when it came to making respectable judgments about this continent's natural history. Allegations of rampant plagiarism and the wholesale fabrication of scientific observations could, as in Constantine Rafinesque's case, cement a reputation for greed that seemed to reinforce an expectation of alien untrustworthiness.

From the start, American sensitivity to European intellectual hegemony was an issue. More than just a holdover from Thomas Jefferson's spirited response to the Comte de Buffon, the steady stream of published accounts by Europeans of their travels in the United States provided a constant reminder that the new nation was being scrutinized and judged by its visitors. Hasty impressions recorded in these widely read books and articles provoked patriotic ire at the least offense. When inaccuracies involved landscape descriptions, American naturalists howled that the general trust in European authority could seriously misdirect geological interpretations and perpetuate fallacious misconceptions about Earth's history. Yale president Timothy Dwight was an early advocate for American self-reliance in geology. Critiquing instances where famous European savants would report hearsay and conjecture as fact, Dwight castigated Constantine François Volney in his essay "Remarks on European Travellers" for completely misrepresenting the configuration of the mountain ranges in northern New England. Dwight added, "He ought not, however, to have asserted so roundly what he did not, and could not, know to be true."[3] By way of contrast, Dwight credited the Duke de la Rochefoucauld's book of travels for "not deriving general conclusions from a single fact, or a very small number of facts; and . . . in shewing no disposition to originate theories."[4] In other words, to report accurate observations was acceptable, but Europeans were courting trouble whenever they ventured further into theoretical interpretations that might "explain" American natural phenomena.

The frontier reputation for placing practical considerations above theoretical ones constitutes an important second trope in the study of early American science. Baconian ideology imbued both the rhetoric and the work of geological investigation in the early national period. New York–based naturalists struggled to cultivate and assert their own expertise in the face of real scientific challenges and perceived cultural barriers. From DeWitt Clinton's celebrated promotion of the state's native grains, birds, and fish as internationally worthy scientific findings to the sensational fossil specimens that Henry Schoolcraft and David Thomas propelled from the canal digger's shovel into popular celebrity, efforts to discover the wonders of New York's natural history combined to produce a palpably nationalistic, yet theoretically insecure set of scientific answers.

A careful review of Eaton's career, however, demonstrates the inadequacy of the truism that historians have repeated about nineteenth-century American science. Eaton's pioneering field research–based modes of surveying and documenting agricultural and mineral resources, his models for effective participatory science education, his impressive compilations of geological profiles, and his mapmaking achievements all look like they confirm the prejudice on behalf of practical results over theoretical speculation. On the other hand, his passionate labors on behalf of a full-blown diluvialist theory and his insistence on developing and promoting the New York system of rock nomenclature can only be understood as clear and consistent attempts to demonstrate that American science could not thrive on practical applications alone. Americans needed to generate and defend their own geological theories, whether that meant direct competition against or friendly collaboration with their European counterparts.

The interactions between theory and practice, as set against the backdrop of cultural aspiration to meet European standards of achievement, become clearer in the light of comparisons to the work of New York's leading writers and painters. Eaton was a personal friend of Washington Irving and a scientific mentor to William Cullen Bryant. In turn, these men were the close friends and intimate colleagues of James Fenimore Cooper and Thomas Cole. The relationship between Eaton's scientific work and the artistic productions of his contemporaries in short fiction, nonfiction expedition accounts, poetry, novels, and landscape painting attests to the broader interdisciplinary significance of geological study in the historical and social contexts of early nineteenth-century New York. At the same time, the struggles to acquire expertise, establish credibility, and secure institutional homes for training and ongoing professional work were not significantly different for individuals working in any of these distinct areas. As Benjamin Silliman and Thomas Brownell had done in the first decade of the century, a highly talented Rensselaer student like the future Harvard professor Eben Horsford would find it necessary to go to Europe to burnish his scientific training after completing his education under Eaton. This kind of seasoning and exposure was not so different from the extended sojourns to Europe that Irving, Cooper, and Cole undertook to nurture their talents and advance their career prospects.

Ultimately, of course, Eaton hoped that New York and the other states would choose to provide secure and meaningful employment for practicing botanists, geologists, mineralogists, and paleontologists. The idea of a state geological survey had begun as a self-promoting entrepreneurial dream. When Eaton returned to Albany after his exile in 1819, he gave lectures to the Society for the Promotion of Agriculture and to the New York State legislature on the virtues and potential rewards of public support for scientific research. Given the chance to test his promises by winning the support of a wealthy private patron, Eaton worked constantly throughout the remaining two dozen years of his life trying to justify and expand Van Rensselaer's investments in geological investigation and education. The larger vision of establishing a systematic network of state natural history surveys work also remained a top priority for Eaton and his colleagues. In the long run Eaton had every right to be proud. New York's example of investing in geological know-how proved highly infectious. By midcentury, twenty-four of the country's thirty state legislatures had enacted and funded geological, mineralogical, or comprehensive natural history surveys. Furthermore, an astounding number of Eaton's former students and protégés found employment as leaders of these surveys, which were the first step toward the development of government-based, professional scientific bureaucracies.[5]

Finally, to reevaluate the metaphor of warfare between science and religion, we must ask how applicable this metaphor is to the history of geology and its repercussions in New York State through the first four decades of the nineteenth century. Conflicts and tension certainly arose as a result of changing scientific modes of understanding the history of the Earth, as well as changing religious modes of answering the same kinds of questions. An important insight to be gained here is not simply that a new science was challenging the authority of an old theology. Throughout the time period examined in this book, new sciences were constantly challenging older scientific traditions and new theologies were challenging older theological traditions. In the midst of all this speculation, interesting and complicated debates erupted over unshared assumptions about the uniformity of the laws of physics across time and space. Could time itself have moved more slowly or quickly in the past than it does in the present? What would constitute persuasive evidence of a terrestrial past that was imaginably different in some key way from the conditions that generally prevail in the present? Was it possible—or even sensible to talk about a past that was unimaginably different? Into which of these categories would you place the workings of a mile-thick ice sheet covering New York State? How about a sudden tsunami-induced, sea-level rise of one thousand feet, followed by the dynamics of gradual drainage? These were, in fact, the kinds of startling, fantastic discussions that the geological record provoked when people began to examine it, whether they did so systematically using the techniques and modes of natural science or just casually using their common sense and the cultural assumptions they had inherited from Judeo-Christian teachings.

Through these sometimes surprising juxtapositions, useful but unexpected patterns can be discerned. With their exploration of physical geography and amid the turbulence of early state and national politics, New Yorkers were forming a dynamic society obsessed with risk and reward. By grasping at the prosperity promised by Enlightenment science, people followed the intellectual interests and avocations of wealthy, powerful patrons and met the strategic and commercial requirements of the early republic's most extensive state in times of war and peace. They also met the material challenges of reshaping the land into a navigable waterway and enjoyed the spinoff benefits of finding valuable fossils and winning scientific prestige. In this way, the essential interdependence of science and technology combined to produce symbolic and theoretical revolutions as well as practical and educational opportunities. Writers and painters tried to invent a rich and meaningful history from the raw materials of an unknown place, while natural philosophers and theologians coped with the anxieties produced by an Earth that was getting older with each new discovery. The struggle between cultural acceptance of and rebellion against technical expertise had begun, and it continues to this day.

Notes

INTRODUCTION: A Meeting Place for Waters and Students of Earth History

1. According to U.S. Census figures, Albany grew dramatically from a population of 5,349 in 1800 to 9,356 in 1810 to 12,630 in 1820; then it almost doubled again to 24,209 in 1830.

2. The second voyage by an American commercial vessel to trade for tea and porcelain in Canton was launched by Albany merchants in 1785. See Paul E. Fontenoy, "An 'Experimental' Voyage to China, 1785–1787," *American Neptune* 55 (1995): 289–300.

3. Edward Countryman, *Americans: A Collision of Histories* (New York: Hill and Wang, 1996), 121.

4. In the special election in the spring of 1817 to replace Gov. Daniel Tompkins, James Monroe's new vice president, Clinton received 43,310 votes. Some 1,479 Tammany Hall–manufactured ballots were cast for Buffalo's Peter B. Porter, but that was Clinton's only opposition.

5. For a detailed rendition of this history of professionalization in American science, see Sally Gregory Kohlstedt, *The Formation of a Scientific Society: The American Association for the Advancement of Science, 1848–1860* (Urbana: University of Illinois Press, 1976).

6. Stephen Van Rensselaer agreed to sponsor the Erie Canal survey after Eaton produced similar reports of Rensselaer's property (Albany and Rensselaer Counties).

7. Simon Winchester's popular book boasts about the remarkable map that William "Strata" Smith produced in 1815, which outlined beds of strata laid out in a three-hundred-mile arc across England and Wales. *The Map That Changed the World: William Smith and the Birth of Modern Geology* (New York: Harper Perennial, 2002). Amos Eaton's achievements by 1830 deserve similar accolades. By combining his own findings with those of Massachusetts state geologist Edward Hitchcock, in 1824 Eaton generated a profile of nearly five hundred miles of continuous geological strata, roughly tracing a line of latitude from Boston to Buffalo. In 1829 he worked to assemble a "three-dimensional" native bedrock map that not only covered the entire state of New York (whose total land area rivals England's) but also displayed two more geological profiles along its margins.

8. Rensselaer School Trustees' Report, 11 Mar. 1828, quoted by Henry B. Nason, *Biographical Record of the Officers and Graduates of the Rensselaer Polytechnic Institute, 1824–1886* (Troy, NY: William H. Young, 1887), 18.

9. George H. Daniels, *American Science in the Age of Jackson* (New York: Columbia University Press, 1968), 11–12.

10. Ibid., 38. See also the adroit discussion of the issue of professionalization in this literature, and Amos Eaton's evocative example in particular, in Paul Lucier, "The Professional and the Scientist in Nineteenth-Century America," *ISIS* 100 (2009): 699–700. For my purposes in this book, the term *professional* reverts to the simpler colloquial meaning, such that a "professional" geologist (or botanist, conchologist, zoologist, etc.) was a naturalist who could earn a living by pursuing scientific activities without having to maintain a separate, full-time occupation outside of the practice or teaching of natural history. Amos Eaton spent the latter half of his life struggling to create a niche for himself as a professional in science in this sense.

11. George Brown Goode, "The Beginnings of American Science: The Third Century," presidential address delivered before the Biological Society of Washington on 22 Jan. 1887; published as part of the collection of Goode's essays entitled *The Origins of Natural Science in America* (1901; repr., with an introduction by Sally Gregory Kohlstedt, Washington, DC: Smithsonian Institutional Press, 1991), 147. According to John Mason Clarke, Eaton and Lesueur spent a day in 1820 hiking together through the Helderberg foothills (on the northern flank of the Catskill mountains) amid a wealth of yet unnamed fossils. Eaton was thus taught about fossil stratigraphy through this direct encounter with "Cuvierian eyes and training." John M. Clarke, *James Hall of Albany: Geologist and Paleontologist, 1811–1898* (Albany, 1923), 89.

12. Michele Aldrich, *The New York State Natural History Survey, 1836–1842* (Ithaca, NY: Paleontological Research Institute, 2000); Paul Lucier, *Scientists and Swindlers: Consulting on Coal and Oil in America, 1820–1890* (Baltimore: Johns Hopkins University Press, 2008).

13. Andrew J. Lewis, *A Democracy of Facts: Natural History in the Early Republic* (Philadelphia: University of Pennsylvania Press, 2011), 139–41.

14. Shocking to the extreme is Keith Thomson's character assassination, beginning with a slew of inaccurate claims and summing up with this caricature: "Eaton was one of the most colorful, if eccentric, often wrong, and sometimes downright obnoxious characters in American science." *The Legacy of the Mastodon: The Golden Age of Fossils in America* (New Haven, CT: Yale University Press, 2008), 101.

15. Art historian Ellwood Parry makes this particular claim in an article addressed to historians of geology, while analyzing images Cole provided to illustrate Englishman John Howard Hinton's *The History and Topography of the United States* (1830/1832). See Ellwood C. Parry III, "Acts of God, Acts of Man: Geological Ideas and the Imaginary Landscapes of Thomas Cole," in *Two Hundred Years of Geology in America*, ed. Cecil Schneer (Hanover, NH: University Press of New England, 1979), 57.

16. See, respectively, Carol Sheriff, *The Artificial River: The Erie Canal and the Paradox of Progress, 1817–1862* (New York: Hill and Wang, 1996); David Stradling, *The Nature of New York: An Environmental History of the Empire State* (Ithaca, NY: Cornell University Press, 2010); and Richard W. Judd, *The Untilled Garden: Natural History and the Spirit of Conservation in America, 1740–1840* (Cambridge: Cambridge University Press, 2009).

17. See Harland Hoge Ballard, "Amos Eaton: A Pioneer of Science in Berkshire County," in *Collections of the Berkshire Historical and Scientific Society* (Pittsfield, MA: Sun Printing Company, 1897), 179–312; and Ethel M. McAllister, "Amos Eaton, Scientist and Educator" (Ph.D. diss., University of Pennsylvania, 1941).

18. Historian Sara S. Gronim argues interestingly that the distinct importance of localized natural knowledge had not been a significant feature of New York's colonial scientific culture: "Shielded by their technology from the need to learn local phenomena deeply, . . . immigrants to New Netherland and New York investigated the region's particularities only superficially." *Everyday Nature: Knowledge of the Natural World in Colonial New York* (New Brunswick, NJ: Rutgers University Press, 2009), 202.

19. Benedict Anderson, *Imagined Communities: Reflections on the Origin and Spread of Nationalism* (London: Verso, 1991); David Copeland, "America: 1750–1820," in *Press, Politics, and the Public Sphere in Europe and North America, 1760–1820*, ed. Hannah Barker and Simon Burrows (Cambridge: Cambridge University Press, 2002), 140–58.

20. See, e.g., Kenneth Pomeranz, *The Great Divergence: China, Europe, and the Making of the Modern World Economy* (Princeton: Princeton University Press, 2000), and H. L. Wesseling, *The European Colonial Empires, 1815–1919* (New York: Longman, 2004).

21. For example, political scientist James W. Ceasar argues that the early republic took a regrettable turn when natural history replaced physics as the dominant scientific model for American political thought. "Fame and *The Federalist*," in *The Noblest Minds: Fame, Honor, and the American Founding*, ed. Peter McNamara (Lanham, MD: Rowman and Littlefield, 1999), 204.

22. Alexis de Tocqueville, *Democracy in America*, trans. George Lawrence (1966; repr., with a preface by J. P. Mayer, Garden City, NY: Doubleday, 1969), 454.

23. Katherine Pandora, "Popular Science in National and Transnational Perspective: Suggestions from the American Context," *ISIS* 100 (2009): 346–58, see esp. 351–53.

CHAPTER 1: Invitations to Study the Earth's Past

Epigraph. X Y Z etc., A *Knickerbocker tour of New York State, 1822*: "Our Travels, Statistical, Geographical, Mineorological, Geological, Historical, Political and Quizzical," ed. with an intro. and notes by Louis Leonard Tucker (Albany: State Education Department, 1968), 110. The lampooned "Mitchell" guidebook was, of course, Samuel Latham Mitchill's.

1. For brief sketches of Buffon's ideas about the age of the Earth, see G. Brent Dalrymple, *The Age of the Earth* (Palo Alto, CA: Stanford University Press, 1991), 29–31, and Pascal Richet's excellent *A Natural History of Time*, trans. John Venerella (Chicago: University of Chicago Press, 2007), 131–34. For a summary of Hutton's controversial ideas and their contemporary reception, see Paolo Rossi's classic work, *The Dark Abyss of Time: The History of the Earth and the History of Nations from Hooke to Vico*, trans. Lydia G. Cochrane (Chicago: University of Chicago Press, 1984), 113–18; Rachel Laudan's thorough discussion in *From Mineralogy to Geology: The Foundations of a Science, 1650–1830* (Chicago: University of Chicago Press, 1987), 113–23; and Richet, *Natural History*, 158–62.

2. Though it is a tale widely repeated among accounts of "first" instances of an idea (the mode that scientists typically invoke as their main interest in history), Agassiz biographer Edward Lurie provides a nicely detailed description of this episode from the Swiss scientist's career. *Louis Agassiz: A Life in Science* (Baltimore: Johns Hopkins University Press, 1988), 94–100.

3. According to George R. Stewart's classic study on place names in the United

States, the name Niagara derived from an early French priest's visit to an Iroquois town near the Lake Ontario outlet of a wide river. The town was called Ongniahhra, which was translated as "point of land cut in two." Stewart remarks that the priest was informed of a waterfall located farther up the river, but he "did not go to look at it, probably thinking all waterfalls are alike." *Names on the Land: A Historical Account of Place Naming in the United States* (New York: Random House, 1945; repr., with an intro. by Matt Weiland, New York: New York Review of Books, 2008), 83.

4. Percy Adams cites Hennepin's *Description de la Louisiane* (Paris, 1683), trans. Marion E. Cross and published as *Father Louis Hennepin's Description of Louisiana* with an intro. by Grace Lee Nute (St. Paul: University of Minnesota Press, 1938) and Hennepin's subsequent *Nouvelle découverte d'un tres grand pays* (Utrecht, 1697), translated and published in England as *New Discovery of a Grand Country Between New Mexico and the Frozen Sea* (London, 1698). Percy G. Adams, *Travelers and Travel Liars, 1660–1800* (1962; repr., New York: Dover Books, 1980), 46–47.

5. Lahontan figures even more culpably than Hennepin in Percy Adams's analysis of the fictions and plagiarism that abounded throughout the early history of North American geography. Although it is unquestioned that Lahontan did spend time at Fort Niagara, his false claim (of having discovered and traveled a thousand miles northwest and back along a great river, the rivière Longue, that empties into the upper Mississippi river somewhere in present-day Minnesota) would tantalize seekers of a water route across North America and plague the imaginations of cartographers for the next 150 years. Adams cites Gilbert Chinard's introduction to Baron de Lahontan, *Dialogues curieux entre l'auteur et un sauvage de bons sens qui a voyagé, et Mémoires de L'Amérique Septentrionale* (Baltimore, 1931); Adams, *Travelers*, 56–57.

6. Peter Kalm, to Carl Christopher Gjörwell, "A Description of Niagara Water Fall in North America," first published in *Gentleman's Magazine* (1751), translated and reprinted as an appendix to *Peter Kalm's Travels in North America* (the English version of 1770), ed. Adolph B. Benson (1937; repr., New York: Dover Books, 1987), 695.

7. Ibid., 696.

8. Ibid., 696–97.

9. Ibid., 698.

10. French linguist, philosopher, and politician Constantin-François Chasseboeuf [Comte de] Volney's geological and climatological study, based on the observations he made during his 1795–98 stay (which was cut short under suspicion of espionage), was published as *A View of the Soil and Climate of the United States of America*, trans. Charles Brockden Brown (Philadelphia: 1804; repr., with an intro. by George W. White, New York: Hafner, 1968), 95. Other late eighteenth- and early nineteenth-century foreign travelers' accounts are too numerous to list, but the range of primary sources includes the observations of a French professional soldier, François-Jean Beauvoir, the Marquis de Chastellux, *Travels in North America in the Years 1780, 1781, and 1782*, 2 vols., trans. and ed. Howard C. Rice Jr. (Chapel Hill: University of North Carolina Press, 1963); François-Alexandre-Frédéric [Duc de] La Rochefoucauld-Liancourt, *Travels through the United States of North America, the country of the Iroquois, and Upper Canada, in the years 1795, 1796, and 1797; with an authentic account of Lower Canada*, 2 vols., trans. Henry Neuman (London: R. Phillips, 1799), which was widely ridiculed by Americans for its abundance of errors; and aristocratic Englishman John Maude's *Visit to the Falls*

of Niagara in 1800 (London: Longmans, Rees, Orme, Brown and Green, 1826; reprinted Louisville, KY: Lost Cause Press, 1980).

11. Chandos Brown's biography remains the best single source on Benjamin Silliman's life and science. Chandos Michael Brown, *Benjamin Silliman: A Life in the Young Republic* (Princeton: Princeton University Press, 1989).

12. Amos Eaton, "The Globe Had a Beginning," *American Journal of Science and the Arts (AJS)* 3 (1821): 238.

13. Edward Hitchcock, paper read before American Geological Society, 11 Sept. 1822, published as "A Sketch of the Geology, Mineralogy, and Scenery of the Regions contiguous to the River Connecticut," *AJS* 6 (1823): 1–86; see esp. 55–56 for quotation.

14. Letter from Andrew Ellicott to Dr. Benjamin Rush, *European Magazine and London Review* (Oct. 1793), excerpted and cited by Catherine Van Cortlandt Mathews, *Andrew Ellicott: His Life and Letters* (New York: Grafton Press, 1908), 79. William Wyckoff's study of the career of Andrew's younger brother Joseph, who assisted in the Niagara frontier survey work, provides a fine-grained analysis of the economic and social conditions under which western New York's landscape was transformed during the transition between Iroquois tribal control and New York State administration. For a summary of contemporary travelers' reports and analysis of the environmental changes then taking place along the Niagara River, see Wyckoff, *The Developer's Frontier: The Making of the Western New York Landscape* (New Haven: Yale University Press, 1988), 152–56.

15. Caleb Atwater to Benjamin Silliman, 28 May 1818 [Circleville, Ohio], published as "On the Prairies and Barrens of the West," *AJS* 1 (1818): 116–25, quotation on 124.

16. Among the many pieces of useful advice to be garnered from William Cronon's excellent work is the suggestion that anyone studying the early ecological history of the northeastern United States should read the four volumes of Timothy Dwight's *Travels* "from cover to cover." William Cronon, *Changes in the Land: Indians, Colonists, and the Ecology of New England* (New York: Hill and Wang, 1983), 210.

17. Historian John Greene characterized Dwight as a sort of "Christian-Federalist counterpart" to Republican deist Thomas Jefferson. John Greene, *American Science in the Age of Jefferson* (Ames: Iowa State University Press, 1984), 14; Thomas Jefferson, *Notes on the State of Virginia* (1785), reprinted in *The Portable Thomas Jefferson*, ed. Merrill D. Peterson (New York: Viking, 1975).

18. Timothy Dwight, *Travels in New-England and New-York*, 4 vols. (New Haven: Timothy Dwight, 1822), 3:204. Herkimer is the township where the Little Falls of the Mohawk are situated. Note that although the letters from Dwight's journeys were composed during the 1790s, 1800s, and 1810s, none were widely known until 1821–22, when posthumous editions of the four volumes of his *Travels* were finally published in New Haven and London. Because of the delayed circumstances of their publication, Dwight's letters fell outside contemporary scientific debates; nevertheless, they provide an authentic window into the contemporary geological imagination.

19. Ibid., 3:351–52.

20. Ibid., 4:132, 183–84.

21. Ibid., 3:445.

22. Ibid., 3:446–47.

23. Ibid., 3:296, 299.

24. Jefferson, *Notes on Virginia*, 48.

25. See J. W. Wilson to Benjamin Silliman, 29 Mar. 1821 [Newburgh], published as "Bursting of Lakes Through Mountains," *AJS* 3 (1821): 252–53, for how all these examples provoked a hostile reader's reaction to the entire theory.

26. Georges Cuvier, *Essay on the Theory of the Earth*, trans. with notes by Robert Jameson (Edinburgh, 1813); repr., ed., and with an appended guide to American geological phenomena by Samuel Latham Mitchill (New York: Kirk and Mercein, 1818).

27. William Darby, *A Tour from the City of new York, to Detroit in the Michigan territory, made between the 2d of May and the 22d of September, 1818* (1819; repr., Chicago: Quadrangle Books, 1962), 48–49.

28. Darby, *A Tour*, xxxviii–xxxix (pages that appear among the addenda following the main text).

29. X Y Z etc., *A Knickerbocker tour of New York State, 1822*, 109–10.

30. Hadassah Davis and Natalie Robinson, *History You Can See: Scenes of Change in Rhode Island, 1790–1910* (Providence: Rhode Island Publications Society, 1986), 58.

31. Dwight, *Travels*, 3:180.

32. John McPhee, *Basin and Range* (New York: Farrar, Straus, Giroux, 1980), 12.

33. Amos Eaton, *A Geological and Agricultural Survey of the district adjoining the Erie Canal, in the State of New York, taken under the direction of the Hon. Stephen Van Rensselaer: Part I. containing a description of the rock formations; together with a geological profile, extending from the Atlantic to Lake Erie* (Albany: Packard and Van Benthuysen, 1824), 8.

CHAPTER 2:   Natural Sciences and Civic Virtues

Epigraph. James Renwick, *Life of Dewitt Clinton* (New York: Harper and Brothers, 1840), 127.

1. This is not to suggest that the roles of politics and science in early republican civic virtue have gone unnoticed. Colleen Terrell, e.g., writes of the nascent Jeffersonian Republican program to harness men of science to advance its credibility in "Republican Machines: Franklin, Rush, and the Manufacture of Civic Virtue in the Early Republic," *Early American Studies* 1 (2003): 120–32. With particular reference to Clinton, biographer Evan Cornog devotes one chapter to the man's intellect, concluding however that the vocation of politics remained absolutely distinct from science, which he characterizes as a mere avocation for Clinton. *The Birth of Empire: DeWitt Clinton and the American Experience, 1769–1828* (New York: Oxford University Press, 1998), 118–26.

2. Quotation is from Gordon S. Wood, *Empire of Liberty: A History of the Early Republic, 1789–1815* (New York: Oxford University Press, 2009), 159. Wood credits Joanne Freeman's work for providing a basis for much of his analysis of honor; see Joanne B. Freeman, *Affairs of Honor: National Politics in the New Republic* (New Haven: Yale University Press, 2001), and "Slander, Poison, Whispers, and Fame: Jefferson's 'Anas' and Political Gossip in the Early Republic," *Journal of the Early Republic* 15 (1995): 25–57.

3. Freeman, *Affairs of Honor*, xvi.

4. James J. Kirschke, *Gouverneur Morris: Author, Statesman, and Man of the World* (New York: St. Martin's, 2005), 163. Kirschke makes this statistical claim but lists just five of the nine colleges, omitting Penn and the three colleges not founded until the waning years of the 1760s.

5. Astronomy and mathematics were the most well-established technical fields within

a classical education, so a course in Newtonian physics was the pinnacle of what learned gentlemen could hope to master when King's was founded in the mid-eighteenth century. Sara S. Gronim, *Everyday Nature: Knowledge of the Natural World in Colonial New York* (New Brunswick, NJ: Rutgers University Press, 2009), 151.

6. Kirschke, *Gouverneur Morris*, 12.

7. This paper is widely regarded as the birth moment for American vertebrate paleontology. In a postscript, Jefferson's Megalonyx (giant lion's claw) paper even cited the contemporary discovery of a similar Paraguayan fossil (which turned out to be the same species) named Megatherium (giant sloth) by a young French naturalist named George Cuvier. Thomas Jefferson, "A Memoir on the Discovery of Certain Bones of a Quadruped of the Clawed King in the Western Parts of Virginia," *Transactions of the American Philosophical Society* 4 (1799): 246–60.

8. Jonathan Daniels describes the 1791 botanizing tour and the reactions it excited in *Ordeal of Ambition: Jefferson, Hamilton, Burr* (New York: Doubleday, 1970), 61–64.

9. Ibid., 104, 110–14.

10. A Layman, "Science and Government," *The Claims of Thomas Jefferson to the Presidency Examined at the Bar of Christianity* (Philadelphia: Dickins, 1800), 49–50, quoted in Edward J. Larson, *A Magnificent Catastrophe: The Tumultuous Election of 1800, America's First Presidential Campaign* (New York: Free Press, 2007), 181.

11. *American Mercury* (31 July 1820), 1; quoted in Larson, *Magnificent Catastrophe*, 182.

12. For examples, see Daniel J. Boorstin, *The Lost World of Thomas Jefferson* (Boston: Beacon, 1948; repr., Chicago: University of Chicago Press, 1993); Edward T. Martin, *Thomas Jefferson, Scientist* (Henry Schuman, 1952; repr., New York: Collier, 1962); Charles A. Miller, *Jefferson and Nature: An Interpretation* (Baltimore: Johns Hopkins University Press, 1988); Silvio A. Bedini, *Thomas Jefferson, Statesman of Science* (New York: Macmillan, 1990); and Lee Alan Dugatkin, *Mr. Jefferson and the Giant Moose: Natural History in Early America* (Chicago: University of Chicago Press, 2009).

13. Samuel L. Mitchill, *A Discourse on the Character and Services of Thomas Jefferson, more especially as a promoter of natural and physical science* (address before the New York Lyceum of Natural History, given 11 Oct. 1826; repr. as *Publications on Thomas Jefferson and Science*, No. 1, Charlottesville: Division of Humanities School of Engineering and Applied Science University of Virginia, 1982), 31.

14. Sara Gronim explains how colonial New York families sometimes invested in university medical studies for their sons as a bid to elevate their social standing. For example, "in 1761 John Bard sent [his son] Samuel to the medical school at the University of Edinburgh, among the most prestigious medical schools in Europe." Gronim, *Everyday Nature*, 126–27.

15. [John] I. Finch, *Travels in the United States of America and Canada, containing some account of their scientific institutions, and a few notices of the geology and mineralogy of those countries, to which is added, An Essay on the Natural Boundaries of Empires* (London: Longman, Rees, Orme, Brown, Green, and Longman, 1833), 56.

16. Washington Irving's biographer Andrew Burstein presents a savagely critical perspective on DeWitt Clinton. The founding of all those learned societies, e.g., represent "Clinton's overly self-conscious attempt to define the contours of Manhattan's intellectual life." *The Original Knickerbocker: The Life of Washington Irving* (New York: Basic Books, 2007), 102.

17. DeWitt Clinton, *An Introductory Discourse Before the Literary and Philosophical Society of New York* (1814; repr., New York: Arno, 1978), 4.

18. Samuel L. Mitchill, *Discourse on the Character and Scientific Attainments of De-Witt Clinton, late Governor of the State of New-York; pronounced at the Lyceum of Natural History, of which he was an honorary member, on the 14 July, 1828* (New York: E. Conrad, 1828), 21.

19. DeWitt Clinton, *An Account of Abimalech Coody and Other Celebrated Writers of New York, in a Letter from a Traveller, to His Friend in South Carolina* (New York: Jan. 1815); quoted in Burstein, *Original Knickerbocker*, 105.

20. Renwick, *Life*, 120. Historian Simon Baatz has written a number of articles on the development of New York's early scientific institutions, and his discussion of the political advantages that accrued to organizations founded by the likes of DeWitt Clinton and Samuel Mitchill, by virtue of their dual prominence as scientists and as politicians, supports my own interpretation. See Simon Baatz, *Knowledge, Culture, and Science in the Metropolis: The New York Academy of Sciences, 1817–1970* (New York: New York Academy of Sciences, 1990), 12–13.

21. Caleb Atwater to Isaiah Thomas, 14 Oct. 1820 [Circleville], folder 2, box A, Caleb Atwater Papers, American Antiquarian Society, Worcester, MA.

22. Atwater to Thomas, 13 Oct. 1820 [Circleville], folder 2, box A, Atwater Papers.

23. Atwater to Thomas, 23 Oct. 1820 [Circleville], folder 2, box A, Atwater Papers.

24. Ibid.

25. Frederick Hall to Benjamin Silliman, 24 Sept. 1813, quoted by Chandos Michael Brown, *Benjamin Silliman: A Life in the Young Republic* (Princeton: Princeton University Press, 1998), 297–98.

26. Clinton, *Introductory Discourse*, 40.

27. Renwick, *Life*, 127.

28. Joseph Ewan, ed., *A Short History of Botany in the United States* (New York: Hafner, 1969), 6.

29. Hibernicus [DeWitt Clinton], *Letters on the Natural History and Internal Resources of the State of New York* (New York: Bliss and White, 1822), 106–7.

30. Ibid., 107.

31. DeWitt Clinton, journal entry, 7 July 1810, in William W. Campbell, ed., *The Life and Writings of DeWitt Clinton* (New York: Baker and Scribner, 1849), 42.

32. Ibid., 43.

33. Hibernicus, *Letters*, 19, 206–7.

34. Mitchill, *Discourse on . . . Clinton*, 21.

35. Hibernicus, *Letters*, 33–34.

36. Ibid., 138, 139.

37. Charlotte M. Porter, *The Eagle's Nest: Natural History and American Ideas, 1812–1842* (Tuscaloosa: University of Alabama Press, 1986), 24.

38. DeWitt Clinton, untitled speech (1794), quoted by William W. Campbell, "Internal Improvements," in Campbell, *Life and Writings*, 23.

39. Finch, *Travels*, 56.

40. Ibid., 56–57.

41. Ibid., 58.

42. None of the candidates in the four-way presidential campaign of 1824 acquired

a clear majority in the Electoral College. To provide for such cases, the Constitution placed the decision into the hands of Congress. Kentucky's Henry Clay offered to throw his supporters to Massachusetts's John Quincy Adams, so that the top popular-vote winner, Tennessee's Andrew Jackson, might not prevail. New York senator Martin Van Buren hoped to broker a compromise for Georgia's William Crawford if Adams failed to secure the majority of states on the first ballot. New York's delegation was evenly divided between Adams and Crawford supporters. Van Rensselaer could not escape Van Buren's politicking, particularly as they shared rooms and meals while in the nation's capital. The pressure mounted as it became more and more clear that the Patroon's vote would effectively determine who would become the nation's sixth president. To cover his dismay at the outcome, Van Buren later published a fanciful story about Van Rensselaer seeing Adams's name first after praying for divine guidance. Henry Clay biographers contend that it was not God but Daniel Webster and Clay who talked Van Rensselaer out of Van Buren's scheme. Adams needed a clear victory in the first ballot or the country would be plunged into another election like that of 1800. See David S. Heidler and Joanne T. Heidler, *Henry Clay: The Essential American* (New York: Random House, 2010), 183–84. Of the four aspirants, both Adams and Clay believed that science was a key to power and prosperity. Barely literate, General Jackson felt that intellectual achievement and federal investment in internal improvement projects were low priorities. Seen in this regard, Van Rensselaer's decision is hardly mysterious.

43. Finch, *Travels*, 58.

44. Walter Richard Wheeler, "'Many new homes have lately been built in this city; all in the modern style . . .': The Introduction of the Gambrel Roof to the Upper Hudson Valley," *Hudson River Valley Review* 21 (2004): 1–11, see esp. 6.

45. Carolyn Merchant, *The Death of Nature: Women, Ecology and the Scientific Revolution* (New York: Harper and Row, 1980), 185; and Francis Bacon, *The New Atlantis* (1626), repr. as part of Henry Morley, ed., *Ideal Commonwealths* (New York: P. F. Collier, 1901), text available at http://oregonstate.edu/instruct/phl302/texts/bacon/atlantis.html.

CHAPTER 3: **The Landlord and the Ex-convict**

Epigraph. Amos Eaton, *Geological Journal D*, 12 Aug. 1824 [Root's Nose], 33–34, 37–38, folder 2, box 2, Amos Eaton Papers, New York State Library, Albany.

1. Julian Ursyn Niemcewicz, *Under their vine and fig tree: travels through America in 1797–1799, 1805, with some further account of life in New Jersey*, trans. and ed. with an intro. and notes by Metchie J. E. Budka (Elizabeth, NJ: Grassmann, 1965), 185; quoted by Martin Bruegel, "Unrest: Manorial Society and Market in the Hudson Valley, 1780–1850," *Journal of American History* 82 (Mar. 1996): 1410–11.

2. Alan Taylor, *The Civil War of 1812: American Citizens, British Subjects, Irish Rebels, and Indian Allies* (New York: Alfred A. Knopf, 2010), 183.

3. Evan Cornog, *The Birth of Empire: DeWitt Clinton and the American Experience, 1769–1828* (New York: Oxford University Press, 1998), 102.

4. Stephen Van Rensselaer to U.S. army superiors (1812), quoted by Brian McKenna, *The War of 1812*, "The Battle of Queenston Heights," accessed Nov. 2002, www.galafilm .com/1812/e/events/queen_backa.html.

5. Joyce Appleby, *Inheriting the Revolution: The First Generation of Americans* (Cambridge: Belknap / Harvard University Press, 2000).

6. W. A. Coffey, *Inside out, or, An interior view of the New-York State Prison: together with biographical sketches of the lives of several of the convicts* (New York: James Costigan, 1823), 15.

7. Coffey, *Inside out*, 15.

8. W. David Lewis, *From Newgate to Dannemora: The Rise of the Penitentiary in New York, 1796–1848* (Ithaca, NY: Cornell University Press, 1965), 31, 35.

9. Amos Eaton to Sarah Cady Eaton, Feb. 7, 1814 [New York]; quoted by Ethel M. McAllister, "Amos Eaton, Scientist and Educator" (Ph.D. diss., University of Pennsylvania, 1941), 147.

10. Harland Hoge Ballard, "Amos Eaton: A Pioneer of Science in Berkshire County," in *Collections of the Berkshire Historical and Scientific Society* (Pittsfield, MA: Sun Printing Company, 1897), 195.

11. Amos Eaton, "A System of Mineralogy: Being the Essential Part of the second edition of Kirwan's Elements of Mineralogy," unpublished manuscript, folder 1, box 3, Eaton Papers.

12. *New York Pardons*, vol. 3 (1811–18), 262, quoted in McAllister, "Amos Eaton," 152.

13. This is among the most frequently cited Eaton quotations to be found in American history scholarship, second only perhaps to: "After the Revolution, a thirst for natural science already seemed to pervade the United States like the progress of an epidemic"; see introduction, note 12). For George Brown Goode, the zeal that Williams students showed for natural history added essential color to his paean to Eaton. Amos Eaton, *Geological Text-Book*, 2nd ed. (Albany: Websters and Skinners, 1832), 16, quoted in George Brown Goode, "The Beginnings of American Science: The Third Century," presidential address delivered before the Biological Society of Washington on 22 Jan. 1887, published as part of the collection of Goode's essays entitled *The Origins of Natural Science in America* (1901; repr., with an intro. by Sally Gregory Kohlstedt, Washington, DC: Smithsonian Institutional Press, 1991), 131. The irascible Dutch Marxist historian of science Dirk Struik, on the other hand, suggested that Eaton's effusive report of student response evidenced some compulsion to build up a portfolio of positive recommendations and advertising copy that might yield him future employment at other colleges or lucrative contracts for public speaking engagements. This interpretation would be more persuasive if Eaton had published the statement prior to 1832. Dirk J. Struik, *Yankee Science in the Making: Science and Engineering in New England from Colonial Times to the Civil War* (New York: Dover Books, 1948/1991), 233. For a straightforward instance of this quotation from Eaton, see Howard Ensign Evans, *Pioneer Naturalists: The Discovery and Naming of North American Plants and Animals* (New York: Henry Holt, 1993), 72.

14. Ballard, "Amos Eaton," 228–29.

15. [Amos Eaton], *A Manual of Botany for the Northern States, by the members of the Botanical class in Williams college* (Albany: Websters and Skinners, 1817), quoted in Ballard, "Amos Eaton," 190.

16. See Amos Eaton, *An Index to the Geology of the Northern States: with a Transverse Section from the Catskill Mountains to the Atlantic* (Leicester, MA: Hori Brown, 1818).

17. Amos Eaton's popular *Manual of Botany for the Northern States* went through eight editions during his lifetime (Albany, 1817). It was reprinted in 1818, 1822, 1824, 1829, 1833, 1836, and 1840.

18. Joseph Ewan, ed., *A Short History of Botany in the United States* (New York: Hafner, 1969), 39.

19. Amos Eaton to John Torrey, 20 Sept. 1817, quoted in McAllister, "Amos Eaton," 180.

20. See McAllister, "Amos Eaton," 181–82; and Gloria Robinson, "Edward Hitchcock," in *Benjamin Silliman and His Circle: Studies on the Influence of Benjamin Silliman on Science in America*, ed. Leonard G. Wilson (New York: Science History Publications, 1979), 52; and H. S. Van Klooster, "Amos Eaton as a Chemist," *Journal of Chemical Education* 15 (1938): 454.

21. Amos Eaton to John Torrey, as quoted in Struik, *Yankee Science*, 262–63; presumably taken from either Andrew D. Rodgers, *John Torrey: A Story of North American Botany* (Princeton: Princeton University Press, 1942), or from McAllister, "Amos Eaton," since Bernard Jaffe's quotation from the same letter is much abbreviated in *Men of Science in America* (New York: Simon and Schuster, 1944), 234.

22. Sally Gregory Kohlstedt, *The Formation of the American Scientific Community* (Urbana: University of Illinois Press, 1976), 10, 18.

23. Caleb Strong, Solomon Williams, Ebenezer Hunt, Josiah Dwight, Elijah H. Mills, and David Hunt, published statement, 24 Nov. 1817 [Northampton], quoted in Ballard, "Amos Eaton," 209.

24. Amos Eaton to Mary Lyon [4 Jan. 1825], Mary Lyon Papers, Five Colleges Archives digital access project, accessed July 2011, http://clio.fivecolleges.edu/mhc/lyon/a/2/ff01/250104/01.htm.

25. Elizabeth Alden Green, *Mary Lyon and Mount Holyoke: Opening the Gates* (Hanover, NH: University Press of New England, 1979), 37.

26. Ewan, *Short History*, 39.

27. Howard Ensign Evans, *Pioneer Naturalists: The Discovery and Naming of North American Plants and Animals* (New York: Henry Holt, 1993), 72.

28. Don Rittner, "Troy's Hart Sisters," *The Mesh--Inside Cyberspace* (Albany, NY: MUG News Service, n.d.), accessed Nov. 2002, www.themesh.com/his78.html. For Eaton's relationship with Almira Phelps, see also Vera Norwood, *Made From This Earth: American Women and Nature* (Chapel Hill: University of North Carolina Press, 1993), 19. Sally Gregory Kohlstedt celebrates Almira Hart Phelps as a textbook writer and early advocate of bringing the education of girls outdoors into nature, but she makes no mention of Eaton's roles either in launching this female educator's career or his advocacy of teaching science in the field. *Teaching Children Science: Hands-On Nature Study in North America, 1890–1930* (Chicago: University of Chicago Press, 2010), 15–18.

29. Constantine Rafinesque would rudely lambaste Amos Eaton's 1818 "Reflections on the History and Structure of the Earth" (an appendix to his first edition of *Index to the Geology of the Northern States*), denying that it should even be called "geology," for "geology describes the world as it is . . . and no one will venture to deny its conclusions, since they arise from facts and existing causes." Eaton's appendix, according to Rafinesque, contained at best "ingenious dreams" based upon "suppositions, conjectures, fictions, presuppositions, probabilities, and plausible causes," all of which should not be considered a part of legitimate Baconian science. A writer for the *North American Review* agreed, dismissing the speculative portion of Eaton's book as its least wise or valuable part and lamenting, "We have had speculative geologists enough." Passages from reviews

by C. S. Rafinesque, *American Monthly Magazine and Critical Review*, and from the uncredited author of the *North American Review*, both quoted in George H. Daniels, *American Science in the Age of Jackson* (New York: Columbia University Press, 1968), 142.

30. Eaton to John Torrey, 14 Aug. 1819, quoted in McAllister, "Amos Eaton," 287.

31. Eaton to Torrey, 22 Nov. 1819, quoted in McAllister, "Amos Eaton," 285.

32. Clinton received 43,310 votes out of 44,789, with no Federalist challenger.

33. *New York Pardons, 1811–1818*, 3:503, quoted in McAllister, "Amos Eaton," 155.

34. McAllister, "Amos Eaton," 187–88.

35. Eaton to Torrey, 29 Sept. 1818, quoted in McAllister, "Amos Eaton," 284.

36. John C. Greene, *American Science in the Age of Jefferson* (Ames: Iowa State University Press, 1984), 101–2.

37. Eaton to Torrey, 20 Feb. 1819, quoted by McAllister, "Amos Eaton," 192. Eaton was fond of the archaic word *shew*, meaning "show."

38. Amos Eaton, *An Index to the Geology of the Northern States with Traverse Sections, Extending from Susquehanna River to the Atlantic, Crossing Catskill Mountains*, 2nd ed. (Troy: William S. Parker, 1820), ix, quoted in Michele L. Aldrich, *New York State Natural History Survey, 1836–1845: A Chapter in the History of American Science* (Ithaca, NY: Paleontological Research Institution, 2000), 8. Eaton refers to William Maclure, *Observations on the Geology of the United States of America: with Some Remarks on the Effect Produced on the nature and Fertility of Soils, by the Decomposition of the Different Classes of Rocks; and an Application to the Fertility of Every State in the Union, in Reference to an Accompanying Geological Map* (Philadelphia: A. Small, 1817), and the first American edition of Georges Cuvier, *Essay on the Theory of the Earth*, ed. Samuel L. Mitchill (New York: Kirk and Mercein, 1818).

39. Eaton to Torrey, 14 Aug. 1819, quoted in McAllister, "Amos Eaton," 284.

40. Eaton to Torrey, 31 May 1819, quoted in McAllister, "Amos Eaton," 290.

41. McAllister minimizes the importance of the title to Eaton, but his dissatisfaction in representing himself any longer as "Mr. Eaton" is transparent enough in the correspondence following his election to the faculty at Castleton. See McAllister, "Amos Eaton," 200.

42. Eaton to Torrey, 24 June 1820; quoted by McAllister, "Amos Eaton," 285.

43. Ibid., 285–86.

44. Eaton to Torrey, 22 Apr. 1824, in response to Torrey to Eaton 15 Apr. 1824, both quoted in McAllister, "Amos Eaton," 292.

45. See Calvin Durfee, *A Sketch of the Life and Services of the Late Professor Amos Eaton* (Boston, 1860); W. J. Youmans, *Pioneers of Science in America* (New York, 1896); George P. Merrill, "Contributions to the History of American Geology," *Report of the United States National Museum for 1904* (Washington, DC: Government Printing Office, 1906), 251–94; George P. Merrill, *The First One Hundred Years of American Geology* (New Haven: Yale University Press, 1924); repr., New York: Hafner, 1969), 75–126; Palmer C. Ricketts, "Amos Eaton, Author, Teacher, Investigator," *Rensselaer Polytechnic Institute Engineering and Science* 45 (1933): 7–17; W. M. Smallwood, "Amos Eaton, Naturalist," *New York History* 18 (1937): 167–88; K. C. Badgely, "Amos Eaton (1776–1842), Inspired Rensselaer," *Northeastern Geology* 4 (1982): 147–50; and Gerald M. Friedman, "Retrospect on Stratigraphy and Sedimentology at Rensselaer Polytechnic Institute," *Compass* 69 (1992): 349–59.

CHAPTER 4: Clinton's Ditch

Epigraph. Thomas Jefferson, *Notes on the State of Virginia* (Paris, 1785); repr., *The Portable Thomas Jefferson*, ed. Merrill D. Peterson (New York: Viking, 1975), 43.

1. New York anti-federalists (like governor George Clinton and judge Robert Yates) were fond of quoting the following passage from Montesquieu's classic *The Spirit of the Laws*:

> It is natural to a republic to have only a small territory, otherwise it cannot long subsist. In a large republic there are men of large fortunes, and consequently of less moderation; there are trusts too great to be placed in any single subject; he has interest of his own; he soon begins to think that he may be happy, great and glorious, by oppressing his fellow citizens; and that he may raise himself to grandeur on the ruins of his country. In a large republic, the public good is sacrificed to a thousand views; it is subordinate to exceptions, and depends on accidents. In a small one, the interest of the public is easier perceived, better understood, and more within the reach of every citizen; abuses are of less extent, and of course are less protected.

Charles-Louis de Secondat, Baron de Montesquieu, *The Spirit of the Laws*, trans. Thomas Nugent (1752), Book VIII, chap. 16, quoted by Brutus [Robert Yates], [anti-federalist essay no. 1], *New York Journal* (18 Oct. 1787), available at www.constitution.org /afp/brutus01.txt.

2. George Washington to François Jean, Marquis de Chastellux, 12 Oct. 1783, which was subsequently rediscovered by Pennsylvania congressman Andrew Stewart in time to influence debate over federal funding of the Chesapeake and Ohio Canal project in 1826. *19th Congress, 1st Session, House Report no. 228, Mr. Stewart's Report on the Chesapeake and Ohio Canal* (Washington: Gales and Seaton, 1826), 2; quoted by James D. Dilts, *The Great Road: The Building of the Baltimore and Ohio, the Nation's First Railroad, 1828–1853* (Palo Alto, CA: Stanford University Press, 1993), 15.

3. Dilts, *Great Road*, 16.

4. Albert Gallatin, *Report of the Secretary of the Treasury on the Subject of Public Roads and Canals* (1808; repr., New York: Augustus M. Kelley, 1968). For a detailed but succinct summary of Gallatin's plan, see John Lauritz Larson, *Internal Improvement: National Public Works and the Promise of Popular Government in the Early United States* (Chapel Hill: University of North Carolina Press, 2001), 59–63.

5. Dilts, *Great Road*, 19.

6. Chandra Mukerji provides an excellent narrative history and sociocultural analysis of the circumstances that enabled people in seventeenth-century France to achieve this marvel. *Impossible Engineering: Technology and Territoriality on the Canal du Midi* (Princeton: Princeton University Press, 2009).

7. Dorothie Bobbe, *DeWitt Clinton* (New York: Minton, Balch, and Co., 1933), 159.

8. Mary Kay Phelan's popularized account of the history of Erie Canal relies heavily on David Hosack, *Memoir of DeWitt Clinton, with Illustrations of the Principal Events of His Life*, 2 vols. (New York, 1829). See Mary Kay Phelan, *Waterway West: The Story of the Erie Canal* (New York: Thomas Y. Crowell Co., 1977), 14–15.

9. Hawley's essays were reprinted in the appendix to Hosack, *Memoir*, 304–41.

10. DeWitt Clinton to Jesse Hawley, 1822, cited in the introduction to Hawley's essays, reprinted in Hosack, *Memoir*, 304.

11. Carol Sheriff characterizes all seven of the original canal commissioners as "wealthy and prominent New Yorkers" whose main interests in serving on the commission seemed to be to represent their various political factions. *The Artificial River: The Erie Canal and the Paradox of Progress* (New York: Hill and Wang, 1996), 19. For more detailed descriptions of each man's background, see Bobbe, *DeWitt Clinton*, 166; Phelan, *Waterway West*, 17–18; and James Renwick, *Life of Dewitt Clinton* (New York: Harper and Brothers, 1840), 164–68.

12. Phelan, *Waterway West*, 28.

13. William L. Stone to David Hosack, 20 Feb. 1829, reprinted in Hosack, *Memoir*, 437.

14. Quotation taken from "Services of Gideon Granger, Myron Holley, John Greig, Nathaniel Howell, and Nathaniel Rochester" reprinted in Hosack, *Memoir*, 429.

15. Patrick Weissend, "The Great Survey," Holland Land Office Museum, accessed Jan. 2003, www2.pcom.net/cinjod/historian/Survey.html.

16. Noble E. Whitford, *History of the Canal System of the State of New York: together with brief histories of the canals of the United States and Canada* (1905; repr., Albany: Brandow, 1906), quote from chap. 2, "Building the Erie," accessed Jan. 2003, www.history.rochester.edu/canal/bib/whitford/1906/chap02.html.

17. Stephen E. Siry, *DeWitt Clinton and the American Political Economy: Sectionalism, Politics, and Republican Ideology, 1787–1828* (New York: Peter Lang, 1990), 204–5.

18. Mary M. Root, "The Building of the Erie Canal," *Backsights* (an online journal about the history of surveying), accessed Jan. 2003, www.surveyhistory.org/the_building_of_the_erie_canal.htm.

19. DeWitt Clinton to Constantine Rafinesque, reprinted as "Notice on the Hydraulic Limestone," *Western Minerva* 1 (1821): 38. For reports by the engineers themselves, see extracts of Benjamin Wright to W. W. Woolsey, 24 June 1820 [Rome] and of Myron Holley to Benjamin Silliman, 20 Jan. 1821 [Albany], reprinted together as "Lime for Water Cement," *AJS* 3 (1821): 230–31.

20. Isaac Briggs to William Darby, 16 Jan. 1819 [Albany], *Annals of Cleveland, 1818–1935*, vol. 1 (1819; repr., Cleveland: Cleveland Works Progress Administration, 1937), 357–58.

21. David Thomas, *Travels Through the Western Country in the Summer of 1816* (1819; repr., Darien, CT: Hafner, 1970), cited in Lydia P. Hecht, "Memoir of David Thomas," Frontenac Historical Society, accessed Jan. 2003, http://members.aol.com/mhecht7725/FRONTENAC/thomas.html. Hecht also provides the quote attributed to Clinton but without citing a source.

22. DeWitt Clinton to David Thomas, 7 Mar. 1820 [Albany], folder 3, box 1, David Thomas Papers, New York State Library.

23. Benjamin Wright to David Thomas, 19 Aug. 1820 [Rome], folder 3, box 1, Thomas Papers.

24. Myron Holley to David Thomas, 8 Feb. 1821 [Albany], folder 3, box 1, Thomas Papers.

25. Clinton to Thomas, 26 Sept. 1821 [Albany], folder 3, box 1, Thomas Papers.

26. Clinton to Thomas, 4 Sept. 1822 [Albany], folder 4, box 1, Thomas Papers.

27. Clinton to Thomas, 25 July 1822 [Albany], 10 Sept. 1822 [Albany], and 4 Sept. 1822 [Albany], folder 4, box 1, Thomas Papers.

28. Amos Eaton to Henry Schoolcraft, Mar. 1820; quoted in William J. Youmans, ed., "Sketch of Amos Eaton," *Popular Science Monthly* 38 (Nov. 1890): 116.

29. Henry R. Schoolcraft to Benjamin Silliman, 5 June 1822 [Buffalo, NY], published as part of "Remarks on the Prints of Human Feet, observed in the secondary limestone of the Mississippi valley," *AJS* 5 (1822): 226 n.

30. Ibid., 226–27 n.

31. Amos Eaton, "Gases, Acids, and Salts, of recent origin and now forming, on and near the Erie Canal; also living Antediluvial Animals," *AJS* 15 (1829): 247–48.

32. Eaton, "Gases, Acids, and Salts," 248, as quoted in *A Book of the United States: Exhibiting Its Geography, Divisions, Constitution and Government,* ed. Grenville Mellen (New York: George Clinton Smith, 1839), 255.

33. Ibid., 249.

34. Amos Eaton to Benjamin Silliman, 12 Mar. 1822 [Troy], published as "On a singular deposit of gravel," *AJS* 5 (1822): 22–23.

35. For an extended discussion of contemporary American scientific discourse regarding these scratches, see David I. Spanagel, "Great Convulsions and Parallel Scratches: The Era of Romantic Geology in Upstate New York," *Northeastern Geology and Environmental Sciences* 17 (1995): 179–82.

36. David Thomas to Professor J. Griscom, 20 Sept. 1829 [Greatfield], excerpted and published as "Diluvial Furrows and Scratches," *AJS* 17 (1830): 408.

37. John and Katherine Palmer Imbrie relate how an appreciation for past glacial activity was first proposed in the 1780s but little discussed by anyone but Alpine mountaineers until skeptical geologists Jean de Charpentier, Louis Agassiz, William Buckland, and Charles Lyell proved to be key converts in the 1830s. *Ice Ages: Solving the Mystery* (Cambridge, MA: Harvard University Press, 1979), 21–28.

38. See William G. Holmes, "Stephen Reed, M.D., and the 'Celebrated' Richmond Boulder Train of Berkshire County, Massachusetts, U.S.A.," *Journal of Glaciology* 6 (1966): 431–37; and Robert H. Silliman, "The Richmond Boulder Trains: *Vera causae* in Nineteenth-Century American Geology," *Earth Sciences History* 10 (1991): 60–72.

CHAPTER 5: Eaton's Agricultural and Geological Surveys

Epigraph. Amos Eaton to Benjamin Silliman, 2 Sept. 1822; quoted in Ethel M. McAllister, "Amos Eaton: Scientist and Educator" (Ph.D. diss., University of Pennsylvania, 1941), 302.

1. John M. Clarke, *James Hall of Albany: Geologist and Paleontologist (1811–1898)* (Albany, 1923), 27.

2. Samuel Latham Mitchill to T. Romeyn Beck, 6 Dec. 1820 [New York], published in *The Papers of Joseph Henry,* vol. 1, *December 1797–October 1832 The Albany Years,* ed. Nathan Reingold (Washington: Smithsonian Institution Press, 1972), 52–53. See also Amos Eaton and T. Romeyn Beck, *A Geological Survey of the County of Albany* (Albany, 1820).

3. Amos Eaton to John Torrey, 21 Feb. 1821, quoted in McAllister, "Amos Eaton," 201.

4. Walter B. Hendrickson, "Nineteenth-Century State Geological Surveys: Early Government Support of Science," in *Science in America Since 1820,* ed. Nathan Reingold (New York: Science History Publications, 1976), 132. See also Amos Eaton, *A Geological and Agricultural Survey of Rensselaer County in the State of New-York to which is annexed a Geological Profile* (Albany: E. and E. Hosford, 1822).

5. Edmund Berkeley and Dorothy Smith Berkeley, *George William Featherston-haugh: The First U.S. Government Geologist* (Tuscaloosa: University of Alabama Press, 1988), 72. According to Eaton's calculations, the Patroon's payment for research under-taken to study the geology of the Erie Canal entailed cumulative expenses in excess of $18,000. Amos Eaton, "Geological Nomenclature, Classes of Rocks, &c.," *AJS* 13 (1828): 360, cited in Henry S. Williams, "The Devonian and Carboniferous Formations of North America," *Bulletin of the United States Geological Survey* 14 (1897): 31.

6. Amos Eaton to Benjamin Silliman, 2 Sept. 1822, quoted by McAllister, "Amos Eaton," 302.

7. Eaton, *Geological Journal A*, 33–34, folder 1, box 2, Amos Eaton Papers, New York State Library, Albany. All subsequent references to Eaton's *Geological Journal*, vols. A (Nov. 1822), B (May–June 1823), C (July–Aug. 1823, Nov. 1823), D (July–Aug. 1824), E (May–June 1826, July 1827), and F (June 1828, May–Oct. 1829), belong to this manuscript collection.

8. Schoolcraft speculated that this tree had at one time been detached from its native forest, translated some distance by a great natural catastrophe, and "suddenly enveloped in a bed of solidifying sand." Henry R. Schoolcraft, "Remarkable Fossil, found about fifty miles S.W. of Lake Michigan, by his Excellency Gov. Lewis Cass and Mr. Henry R. Schoolcraft, in August 1821, on the River Des Plaines, in the N.E. angle of the State of Illinois, extracted from a paper presented by Mr. Schoolcraft to the American Geologi-cal Society," *AJS* 4 (1822): 287. For the testimonials composed by presidents John Adams, Jefferson, and Madison, see "Honorable Notice of Mr. Schoolcraft's memoir of a Fossil tree," *AJS* 5 (1822): 23–25. I suspect that no such prestigious combination of lay co-authors has since graced the pages of any peer-reviewed American scientific publication.

9. Alexander Macomb was an opportunistic merchant and land speculator who pur-chased almost 3.7 million acres of land in 1791 at twelve cents per acre. His holdings, formerly the homelands of Iroquois tribes that had sided with the British in the Revo-lutionary War, stretched from the eastern tip of Lake Ontario eastward along the Saint Lawrence River and south of the forty-fifth parallel to comprise virtually all four of the counties flanking New York's Canadian boundary on the north and the mountainous regions on the south. Macomb's fortunes turned sour after the Panic of 1792 exposed his inability to pay the property taxes on this vast tract. (Note that the State of New York's extensive Adirondack Park consists of just 2.5 million acres.) After spending a brief period in debtors' prison and being forced to resell much of his land at the discounted rate of one cent per acre, Macomb never regained his fortune. Macomb's son Alexander Jr., however, salvaged the family's respectability in 1814 by successfully defending the north-ern outpost of Plattsburgh, New York, against a combination of British ground and naval forces in the Battle of Lake Champlain. Mount Macomb, the southernmost of the high peaks in the Dix range of the Adirondacks, was named in the younger Macomb's honor after the War of 1812. He was awarded the Congressional Gold Medal for his services in the war, was an active member of Washington, D.C., learned societies, and eventually was promoted to the most lofty post in the army, serving as commanding general of the U.S. Army from 1828 to 1841. David B. Dill Jr., "Macomb's Years in New York City: Wealth and Power," *Watertown Daily Times* (16 Sept. 1990), C1–C2, and "The Audacity of Macomb's Purchase," *Watertown Daily Times* (23 Sept. 1990), C1–C2, C9.

10. Eaton, *Geological Journal C*, 19–22.

11. Bradford B. Van Diver, *Roadside Geology of New York* (Missoula, MT: Mountain Press, 1985), 202–3.

12. Eaton, *Geological Journal C*, 42.

13. Eaton, *A Geological and Agricultural Survey of the Erie Canal*, 101.

14. Eaton, *Geological Journal C*, 61–63.

15. Eaton, *A Geological and Agricultural Survey of the Erie Canal*, 105–6.

16. James Geddes, "Observations on the Geological Features of the South Side of the Ontario Valley," *AJS* 11 (1826): 213–18. See also Robert Bakewell, *An Introduction to Geology: Intended to Convey a Practical Knowledge of the Science, and comprising the most recent discoveries, with explanations of the facts and phenomena which serve to confirm or invalidate various geological theories*, ed. Benjamin Silliman (New Haven: Hezekiah Howe, 1833).

17. Stephen Van Rensselaer to Amos Eaton, 30 Aug. 1823, folder 3, box 1, Eaton Papers.

18. Van Rensselaer to Eaton, 1 Oct. 1823, folder 3, box 1, Eaton Papers.

19. Eaton, *A Geological and Agricultural Survey of the Erie Canal*, 10.

20. For a discussion of the intense wrangling this vacancy triggered among New York's congressional delegation, culminating in a discussion by president James Monroe's cabinet, see Noble E. Cunningham Jr., *The Presidency of James Monroe* (Lawrence: University Press of Kansas, 1996), 122.

21. Van Rensselaer to Eaton, 10 Feb. 1824 [Washington], folder 3, box 1, Eaton Papers.

22. Eaton, *A Geological and Agricultural Survey of the Erie Canal*, 5.

23. McAllister, "Amos Eaton," 276. The present whereabouts of these collections of specimens is unknown. Gerald Friedman has inquired regarding the status of the "Rocks of North America" donation in the name of Stephen Van Rensselaer to the Geological Society at London, 2 July 1825. According to the society representative at Burlington House, these materials were transferred to the Natural History Museum on Cromwell Road in London. Gerald M. Friedman to J[ohn] C. Thackray, 19 Apr. 1994 (copy furnished to author).

24. Paul Lucier distinguishes Eaton's personal service for Van Rensselaer from subsequent, more purely commercial work done by geologists who consulted for the mining industry. "Commercial Interests and Scientific Disinterestedness: Consulting Geologists in Antebellum America," *ISIS* 86 (1995): 248–49.

25. This is how Eaton referred to Amherst College in his letter to John Torrey, 8 Feb. 1823, quoted in McAllister, "Amos Eaton," 208.

26. Amos Eaton, *Geological Journal D*, 24 May 1824 [Troy and Albany], 1. (This is page 1 of a new pagination sequence commencing his fifth canal tour, the journal of which is bound together in the same volume with journals of the third and fourth tours.)

27. William Buckland, "Account of an assemblage of fossil teeth and bones of an elephant, rhinoceros, hippopotamus, bear, tiger, and hyaena, and sixteen other animals, discovered in a cave at Kirkdale, Yorkshire, in the year 1821," *Philosophical Transactions of the Royal Society of London* (1822): 171–236. An excerpt of this paper is reproduced in Martin J. S. Rudwick, *Scenes From Deep Time: Early Historical Representations of the Prehistoric World* (Chicago: University of Chicago Press, 1992), 38–40.

28. William Buckland, *Reliquiae Diluvianae: or, observations on the organic remains contained in caves, fissures, and diluvial gravel, and on other geological phenomena, attesting to the action of an universal deluge* (London: John Murray, 1823). For a descrip-

tion of the Kirkdale cave discoveries and their long-term implications for diluvialism, see Nicolaas A. Rupke, *The Great Chain of History: William Buckland and the English School of Geology, 1814–1849* (Oxford: Clarendon Press, 1983), 32–33, 39–40. Rupke also explains how Lyell's uniformitarian challenge had to deal with well-established practices of historical geology. See Rupke, *Great Chain*, 193–200. Finally, see Charles Lyell, *Principles of Geology; Being an Attempt to explain the Former Changes of the Earth's Surface, by Reference to Causes Now in Operation*, 3 vols. (London: John Murray, 1830–33; ed. and with a new intro. by Martin J. S. Rudwick, Chicago: University of Chicago Press, 1990).

29. [Edward Hitchcock], "Notice and Review of the *Reliquiae Diluvianae: or observations on the organic remains contained in caves, fissures, and diluvial gravel, and on other geological phenomena attesting to the action of an universal deluge*, by the Rev. William Buckland," *AJS* 8 (1824): 150–68, 317–38.

30. Amos Eaton, *Geological Text-Book, prepared for popular lectures on North American Geology; with applications to agriculture and the arts* (Albany: Websters and Skinners, 1830), 14.

31. Mary Kay Phelan, *Waterway West: The Story of the Erie Canal* (New York: Thomas Y. Crowell Co., 1977), 48.

32. Eaton, *Geological Journal D*, 7 June 1824 [Utica], 4–5.

33. Eaton recorded the list of topics upon which he lectured, including chemistry, natural history, galvanism, zoology, phrenology, geology, and botany. From 9 June until 19 July 1824, he lectured in at least one of the cities each day, sometimes delivering two distinct lectures in Rome and Utica on the same day. The only exception to this grueling schedule was a gap of two days (27–28 June) when inflamed lungs and a fever got to be too much for him. See Eaton, *Geological Journal D*, 16 June–12 July 1824 [Rome and Utica], 6–13.

34. Ibid., 6–8.

35. Ibid., 9.

36. Ibid., 11–13.

37. Ibid., 4 Aug. 1824 [Cayuga Lake], 15–16.

38. Ibid., 4 Aug. 1824 [Cayuga Lake], 16–18. Of course, it would be anachronistic and "Whiggish" to suppose that Eaton was anticipating the glacial hypothesis as such.

39. Many thanks to my colleague Josh Rubinstein and Chuck Porter (editor of the *Northeastern Caver*) for their helpful comments and modern descriptions of this cave, which is still designated Mitchell's Cave because its location was first mentioned by Samuel Latham Mitchill. My knowledge about Amos Eaton's team's entry into the cave in 1824, in turn, prompted the publication of some brief excerpts from my primary source notes; see David Spanagel, "Mitchells Cave in 1824," *Northeastern Caver* 32 (2001): 55–56.

40. Eaton, *Geological Journal D*, 12 Aug. 1824 [Root's Nose], 33–34, 37–38.

41. For an institutional history of the Rensselaer Polytechnic Institute, see Samuel Rezneck, *Education for a Technological Society: A Sesquicentennial History of Rensselaer Polytechnic Institute* (Troy, NY, 1968).

42. Michele L. Aldrich, *New York State Natural History Survey, 1836–1845: A Chapter in the History of American Science* (Ithaca, NY: Paleontological Research Institution, 2000), 14.

43. Though he would eventually follow in his family's political footsteps and serve as mayor of Buffalo, New York, George Washington Clinton was initially interested in

natural history and the study of medicine. The governor's son numbered among the first group of assistants who accompanied Eaton's first Rensselaer School field trip along the Erie Canal in 1826. Cortlandt Van Rensselaer, a future Presbyterian clergyman, served in the spring of 1829 as companion to Eaton on research expeditions specifically aimed at securing details for his geological map of New York's strata; to represent portions of the state outside of the well-traveled canal corridor, Eaton needed to examine geological features approaching the state's New England and Pennsylvania frontiers.

44. Stephen Van Rensselaer to Samuel Blatchford, 5 Nov. 1824. Ethel McAllister persuasively asserts that Eaton drafted this letter of intent and incorporation, which the Patroon sent to the institute's first designated president. For the complete text of the letter, see McAllister, "Amos Eaton," 368–71.

45. Carol Sheriff, e.g., cites the Rensselaer Institute, founded in 1824, as the first American school of civil engineering when she makes the conventional observation that the Erie Canal had had to be designed and built by people who were learning engineering on the job. *The Artificial River: The Erie Canal and the Paradox of Progress* (New York: Hill and Wang, 1996), 36.

CHAPTER 6: Empire State Exports

Epigraph. Amos Eaton, *Geological Text-Book, prepared for popular lectures on North American Geology; with applications to agriculture and the arts* (Albany, Websters and Skinners, 1830), vi.

1. Details of the celebration come from the contemporary memoir by Cadwallader D. Colden, *Memoir, Prepared at the Request of a Committee of the Common Council of the City of New York* (New York: W. A. Davies, 1825); quoted in Mary Kay Phelan, *Waterway West: The Story of the Erie Canal* (New York: Thomas Y. Crowell Co., 1977), 100.

2. For landmark works along these lines, see Whitney Cross, *The Burned-Over District: The Social and Intellectual History of Enthusiastic Religion in Western New York, 1800–1850* (Ithaca, NY: Cornell University Press, 1950); Paul E. Johnson, *A Shopkeeper's Millennium: Society and Revivals in Rochester, New York, 1815–1837* (New York: Hill and Wang, 1979); and Michael Barkun, *Crucible of the Millennium: Burned-Over District of New York in the 1840s* (Syracuse, NY: Syracuse University Press, 1986).

3. Amos Eaton to Benjamin Silliman, 23 Nov. 1826, published as "Notices respecting Diluvial Deposits in the State of New-York and elsewhere," *AJS* 12 (1827): 18.

4. Eaton, "Notices respecting Diluvial Deposits," 19.

5. Eaton, *Geological Journal E*, 26 Aug. 1826 [Troy], 46–48, folder 3, box 2, Eaton Papers. Note: a copy of Eaton's *Geological Journal G* can be found at the Rensselaer Polytechnic Institute library archives.

6. Eaton, *Geological Journal E*, 26 Aug. 1826 [Troy], 48–50.

7. Eaton, *Geological Journal E*, 13 Nov. 1826 [Troy], 55.

8. See John M. Clarke, *James Hall of Albany: Geologist and Paleontologist (1811–1898)* (Albany, 1923), 30–35; and Gerald M. Friedman, "Geology at Rensselaer: A Historical Perspective," in *Guidebook [to the] Joint Annual Meeting of the New York State Geological Association . . . and [the] New England Intercollegiate Geological Conference* (held in Troy, New York, on 5–7 Oct. 1979), 5.

9. H. S. Van Klooster, "Amos Eaton as a Chemist," *Journal of Chemical Education* 15 (1938): 458.

10. E. N. Horsford, "Proceedings of the semicentennial celebration of the Rensselaer Polytechnic Institute" (Troy, NY, 1875), quoted in Van Klooster, "Amos Eaton," 458.

11. Amos Eaton, *Prospectus of Rensselaer School* (1827), quoted in Ethel M. McAllister, "Amos Eaton: Scientist and Educator" (Ph.D. diss., University of Pennsylvania, 1941), 390. Heinrich Pestalozzi was another European theorist whose comparable rejection of rote learning would be enthusiastically embraced by educational reformer Horace Mann and the geologist and utopian communitarian William Maclure. See Sally Gregory Kohlstedt, *Teaching Children Science: Hands-On Nature Study in North America, 1890–1930* (Chicago: University of Chicago Press, 2010), 27–28.

12. See Samuel Rezneck, "A Traveling School of Science on the Erie Canal in 1826," *New York History* 40 (1959): 255–69.

13. Eaton, *Geological Journal E*, 2 May 1826 [West Troy], 3. For more information on the subsequent careers of some of the members listed in the expedition, see *The Papers of Joseph Henry*, vol. 1, *December 1797–October 1832 The Albany Years*, ed. Nathan Reingold (Washington: Smithsonian Institution Press, 1972), 135 n.

14. John M. Clarke, "The reincarnation of James Eights, Antarctic Explorer," *Scientific Monthly* 2 (Feb. 1916): 189.

15. See Joseph Henry, "Notes of a Tour from the Hudson to Lake Erie in May and June of 1826," *Papers of Joseph Henry*, 1:136.

16. Eaton, *Geological Journal E*, 8 May 1826 [Whitesborough, Oriskany, Rome, Chittenango], 9. It is indicative of the flux in Eaton's ideas that in the first ten pages of the journal notes for this tour, wherever the words "diluvial gravel" were initially written, Eaton subsequently crossed them out and wrote "crag" above them.

17. Ibid., 15, 13.

18. Ibid., 30 May 1826 [Brighton, Pittsford], 27–28.

19. Joseph Henry, "Canal Tour Journal," 30 May 1826, *Papers of Joseph Henry*, 1:153.

20. Asa Fitch diary, quoted by Rezneck, "A Traveling School," and by Ronald E. Shaw, *Canals For a Nation: The Canal Era in the United States, 1790–1860* (Lexington: University Press of Kentucky, 1990), 192.

21. Caleb Atwater to Isaiah Thomas, 29 Sept. 1820 [Circleville, Ohio], folder 2, box A, Atwater Papers, American Antiquarian Society, Worcester, Massachusetts.

22. Atwater to DeWitt Clinton, 2 Feb. 1823 [Circleville], folder 233, box 4, Henry Post Papers, New York State Library, Albany, NY.

23. Atwater to Henry Post, 3 Jan. 1825 [Circleville], folder 277, box 4, Henry Post Papers.

24. Atwater to Post, 12 Mar. 1828 [Circleville], folder 218, box 4, Henry Post Papers.

25. Joshua V. H. Clark, "A Biographical Sketch of the Hon. James Geddes," *Onondaga; or Reminiscences of Earlier and Later Times*, vol. 2 (Syracuse, NY: Stoddard and Babcock, 1849), 45–50; and Wendy J. Adkins, "Ohio Canals," 1997, www.geocities.com/Heartland/Prairie/6687/.

26. DeWitt Clinton to Micajah T. Williams, 8 Nov. 1823 [New York], published as "His Communication to the Canal Commissioners of the State of Ohio Relative to the Ohio Canal," reprinted in the appendix to David Hosack, *Memoir of DeWitt Clinton, with illustrations of the Principal Events of His Life*, vol. 2 (New York, 1829).

27. Adkins, "Ohio Canals."

28. Wendy J. Adkins, "Ohio and Erie," 1997, www.geocities.com/Heartland/Prairie/6687/ohio.htm.

NOTES TO PAGES 130–136   239

29. DeWitt Clinton, "[Letter] to George P. M'Culloch, Charles Kinsey of Essex, and Thomas Capner, Esqrs. Commissioners of the state of New-Jersey, in relation to a canal from the Delaware to the Passaic," 24 Oct. 1823 [New York], reprinted in the appendix to Hosack, *Memoir*.

30. DeWitt Clinton, "[Letter] to Hon. C.D. Colden, President of the Morris Canal Company," 19 May 1827 [New York], reprinted in the appendix to Hosack, *Memoir*.

31. DeWitt Clinton, "Governor Clinton's Observations relative to the Hampshire and Hampden Canals," preface to letter, "To Samuel Hinckley, James Hillhouse, and Thomas Sheldon, Esqrs. a committee of the Hampshire and Hampden Canal Company," 18 Jan. 1828 [Albany], reprinted in the appendix to Hosack, *Memoir*.

32. DeWitt Clinton, "Governor Clinton's Observations relative to the proposed Delaware and Raritan Canal," preface to letter, "To a gentleman in New-Jersey," 22 Jan. 1828 [Albany]; reprinted in the appendix to Hosack, *Memoir*.

33. John Lauritz Larson does a wonderful job analyzing why transcontinental railroads eventually got built in North America; it only happened when government provided the incentives necessary to overcome initial barriers of financial risk, while promising to allow unregulated market forces to serve private corporate interests once technical success was assured. *Internal Improvement: National Public Works and the Promise of Popular Government in the Early United States* (Chapel Hill: University of North Carolina Press, 2001).

34. Carol Sheriff describes the zero sum game played by upstate New York merchant Lyman Spalding in relocating his grocery business from the turnpike-based Canandaigua to the canal-created Lockport in 1822. *The Artificial River: The Erie Canal and the Paradox of Progress, 1817–1862* (New York: Hill and Wang, 1996), 116–17.

35. Andro Linklater, *Measuring America: How an Untamed Wilderness Shaped the United States and Fulfilled the Promise of Democracy* (New York: Walker and Co., 2002), 181.

36. Table of Contents, *AJS* 14 (1828): 1.

37. Jeremias Van Rensselaer, *Essay on Salt Containing Notices of Its Origin, Formation, Geological Position and Principal Localities* (New York: O. Wilder and J. M. Campbell, 1823).

38. Jeremias Van Rensselaer, *Lectures on Geology: being outlines of the science delivered in the New York Athenaeum in the year 1825* (New York: E. Bliss and E. White, 1825); claim on its behalf made by Gerald M. Friedman, "Retrospect on Stratigraphy and Sedimentology at Rensselaer Polytechnic Institute," *Compass* 69 (1992): 351.

39. Jer[emias] Van Rensselaer, M.D., "Notice of a recent discovery of the fossil remains of the mastodon, *AJS* 11 (1826): 246–50.

40. Amos Eaton, *A Geological and Agricultural Survey of the district adjoining the Erie Canal in the state of New York, taken under the direction of the Hon. Stephen Van Rensselaer: Part I* (Albany: Packard and Van Benthuysen, 1824), 6.

41. Eaton, *A Geological and Agricultural Survey*, 7.

42. Stephen Van Rensselaer to Amos Eaton, 30 Aug. 1822, folder 3, box 4, Eaton Papers.

43. Eaton notes in a journal entry: "This day, Mrs. Eaton brings me this book, which had been lost five months, and I had despaired of ever finding it. I was apprehensive that much of this tour must be reviewed." *Geological Journal B*, 17 Nov. 1823 [Troy], 43.

44. Newell, review of *George William Featherstonhaugh*, by Edmund Berkeley and Dorothy Smith Berkeley, *Journal of the Early Republic* 9 (1989): 263.

45. See Edmund Berkeley and Dorothy Smith Berkeley, *George William Featherstonhaugh: The First U.S. Government Geologist* (Tuscaloosa: University of Alabama Press, 1988), 228–70.

46. Ibid., 11, 26, 27.

47. Ibid.,30–32.

48. Eaton to Benjamin Silliman, 18 May 1829, quoted in McAllister, "Amos Eaton," 294.

49. Nicolaas A. Rupke, *The Great Chain of History* (Oxford: Clarendon Press, 1983), 127.

50. The attacks crested in an exceedingly savage review published after Eaton's *Geological Prodromus* came out. George W. Featherstonhaugh, "Geology," *North American Review* 33 (1831): 471. George Daniels retells the episode, noting how harsh Featherstonhaugh's review was but also remarking with some surprise that the attack centered on Eaton's assertion of an innovative nomenclature rather than his "obviously unscientific" and almost mystical embrace of the number five in nature. See Amos Eaton, "The Number Five, The Most Favorite Number of Nature," *American Journal of Science* 16 (1829): 172–73, as discussed in George H. Daniels, *American Science in the Age of Jackson* (New York: Columbia University Press, 1968), 175–76. In another example, Charlotte Porter regards the same 1831 attack by Featherstonhaugh as being representative of the kind of derision Eaton's ideas suffered, and then she curiously "balances" this notoriously negative review by alluding to "a recent reevaluation" by George Perkins Merrill. Merrill's *The First One Hundred Years of American Geology* was published in 1924. Charlotte M. Porter, *The Eagle's Nest: Natural History and American Ideas, 1812–1842* (Tuscaloosa: University of Alabama Press, 1986), 74 (also see note 5 on 293).

51. George W. Featherstonhaugh, "On the Order of Succession of the Rocks Composing the Crust of the Earth," *Monthly American Journal of Geology and Natural Science* 1 (1832): 337–47; as discussed by Henry S. Williams, "The Devonian and Carboniferous Formations of North America," *Bulletin of the United States Geological Survey* 14 (1897): 35.

52. Amos Eaton to G. W. Featherstonhaugh, Esq., 12 June 1829 [New York], copied into Eaton, *Geological Journal F*, 12 June 1829.

53. Eaton, *Geological Journal F*, 24 June 1829 [Troy], unpaginated.

54. Ibid.

55. Amos Eaton, *Geological Text-Book, prepared for popular lectures on North American Geology; with applications to agriculture and the arts* (Albany, Websters and Skinners, 1830), vi.

56. Ibid., iii.

57. Amos Eaton, *Eaton's Geological Note Book, for the Troy Class of 1841*, 1, folder 5a, box 6, Eaton Papers.

58. James Hall, *New York State Natural History Survey. Geology of New York*, part IV, *Comprising the Survey of the Fourth Geological District* (Albany, 1843), 6; quoted in John C. Greene, *American Science in the Age of Jefferson* (Ames: Iowa State University Press, 1984), 251.

59. Clarke, *James Hall*, 38–39.

60. John M. Clarke and Charles Schuchert, "The Nomenclature of the New York Series of Geological Formations," *Science* 10 (1899): 875.

61. Rachel Laudan lists Eaton's "Tabular View of North American Rocks" (1828) among the pioneering efforts to generate a standard ordering of the Earth's geological formations. This international list included works such as Alexander von Humboldt, *Geognostical Essay on the Superposition of Rocks* (1823); Henri de la Beche, "Synoptical Table of Equivalent Formations" (1824); Ami Boué, "Synoptical Table of the Formations of the Crust of the Earth" (1825); and Alexandre Brongniart, *Tableaux de terrains qui composent l'ecorce du globe* (1829). Rachel Laudan, *From Mineralogy to Geology: The Foundations of a Science, 1650–1830* (Chicago: University of Chicago Press, 1987), 159–61.

62. Michele L. Aldrich, *New York State Natural History Survey, 1836–1845: A Chapter in the History of American Science* (Ithaca, NY: Paleontological Research Institution, 2000), 40. Aldrich cites William Meade, "Chemical Analysis and Description of the Coal Lately Discovered Near the Tioga River, in the State of Pennsylvania," *AJS* 13 (1828): 32–35.

63. As Henry Williams would conclude seven decades later: "This error of Eaton's in identifying the rocks of Ithaca, Cayuga Lake and westward to Lake Erie with the 'Third Grauwacke,' placing them above the Blossburg coal of Pennsylvania, was not corrected until several years later, when the study of fossils clearly revealed the fact that the rocks belonged below the Carboniferous." Williams, "The Devonian," 33.

64. Aldrich, *New York State*, 42.

65. Ibid.

66. Eaton, *Geological Text-Book* (1830), 17.

67. Amos Eaton, *Geological Text-Book, for aiding the study of North American Geology: being a systematic arrangement of facts, collected by the author and his pupils, under the patronage of the Hon. Stephen Van Rensselaer* (Albany: Webster and Skinners, 1832), 3.

CHAPTER 7: Literary Naturalists

Epigraph. Samuel Latham Mitchill, *A Discourse on the State and Prospects of American Literature delivered at Schenectady, July 24, 1821* (Albany, NY: Websters and Skinners, 1821), 33–34.

1. Russel Blaine Nye makes this point explicitly in his classic work of intellectual history, *The Cultural Life of the New Nation* (New York: Harper and Row, 1960), 240.

2. J. Hector St. John de Crèvecoeur, *Letters from an American Farmer and Sketches of Eighteenth-Century America* (first published 1782; repr., with an intro. by Albert E. Stone, New York: Penguin, 1981), 44.

3. Washington Irving to Amos Eaton, 15 Dec. 1802 [New York], in the Simon Gratz Papers, Historical Society of Pennsylvania, Philadelphia, quoted by Ethel M. McAllister, "Amos Eaton: Scientist and Educator" (Ph.D. diss., University of Pennsylvania, 1941), 22–23. On the margin of the letter, regarding the object of Irving's affections, Eaton wrote, "Keep this a secret–he confides in me–it is Matilda Hoffman." The girl was just eleven years old at the time.

4. Thomas Jefferson would appoint Judge Livingston to an opening on the U.S. Supreme Court in 1806.

5. Biographical details on the Irving siblings are nicely outlined in Andrew Burstein, *The Original Knickerbocker: The Life of Washington Irving* (New York: Basic Books, 2007), 12–20, 27–28, quotation from 20.

6. Burstein, *Original Knickerbocker*, 30.

7. [Washington Irving], *Salmagundi; or, The Whim-Whams and Opinions of Launce-lot Langstaff, Esq., and Others* (13 Feb. 1807), in *Washington Irving: History, Tales, and Sketches* (New York: Literary Classics of the United States, 1983), 81.

8. Washington Irving to William Irving (autumn 1814), quoted in Burstein, *Original Knickerbocker*, 100. Burstein provides this quotation (as fully reproduced here) in his text but fails to cite the source of the original manuscript holding.

9. DeWitt Clinton, *An Account of Abimelech Coody and Other Celebrated Writers of New York, in a Letter from a Traveller, to His Friend in South Carolina* (New York: Jan. 1815), 3, 8, quoted by Burstein, *Original Knickerbocker*, 104.

10. Diedrich Knickerbocker [Washington Irving], *A History of New York* (1809), in *Washington Irving: History, Tales, and Sketches* (New York: Literary Classics of the United States, 1983), 394–96.

11. William L. Hedges, *Washington Irving: An American Study, 1802–1832* (Baltimore: Johns Hopkins University Press, 1965), 35.

12. Washington Irving, "The Legend of Sleepy Hollow" (1819), in *Washington Irving: History, Tales, and Sketches*, 1058–88, quoted passages from 1058 and 1075.

13. Washington Irving, "A Virtuoso" from chapter 1 of *A Tour on the Prairies* (Philadelphia: Carey, Lea and Blanchard, 1835), http://etext.lib.virginia.edu/toc/modeng/public/IrvTour.html.

14. Washington Irving, *A Tour on the Prairies* (1835), in vol. 9 of *The Works of Washington Irving* (New York: George P. Putnam, 1851), 161–62.

15. Charles Lyell's landmark *Principles of Geology* was first published in three volumes between 1830 and 1833.

16. Washington Irving, *Astoria* (1836), in vol. 2 of *The Works of Washington Irving* (New York: George P. Putnam, 1849), 179. Incidentally, Irving's use of the word *enthusiasm* deserves some comment here. As Matthew Pethers observes, Enlightenment thinkers took rather a dim view of this term, which emerged directly into American usage as a derogation of the overwrought emotional state attributed to religious zealots inspired by the Great Awakening of the 1790s. "'Balloon Madness': Politics, Public Entertainment, the Transatlantic Science of Flight, and Late Eighteenth-Century America," *History of Science* 48 (2010): 203.

17. See H. M. Brackenridge, *Journal of a voyage up the river Missouri, performed in 1811* (Baltimore, 1816), in vol. 7 of *Early Western Travels, 1748–1846*, ed. Reuben Gold Thwaits (Cleveland, OH: Arthur H. Clark Co., 1904), 102. Jeannette E. Graustein discusses Irving's elaboration of the Brackenridge account and notes the great professional discomfort it caused Nuttall just as he was laboring to analyze results of a recent expedition to Oregon in 1836. *Thomas Nuttall Naturalist: Explorations in America, 1808–1841* (Cambridge, MA: Harvard University Press, 1967), 325–36.

18. Thomas Nuttall, "Observations on the Geological Structure of the Valley of the Mississippi," *AJS* 2 (1821): 14–52.

19. Irving, *Astoria*, 217.

20. Ibid., 218.

21. Washington Irving, *The Adventures of Captain Bonneville* (1837), in vol. 2 of *The Works of Washington Irving* (New York: George P. Putnam, 1849), 40.

22. Irving, *Bonneville*, 160.

23. Joseph Henry, entry from canal tour journal, 27 May 1826, in vol. 1 of *The Papers of Joseph Henry*, ed. Nathan Reingold (Washington: Smithsonian Institution Press, 1972), 152.

24. Hedges, *Washington Irving*, 91. Allusion is obviously made to Bryant's magnum opus, the poem "Thanatopsis." Specific references are also made to Bryant's poem "The Prairies" and to chapter 22 of James Fenimore Cooper's novel *The Prairie*.

25. Gilbert H. Muller, *William Cullen Bryant: Author of America* (Albany: SUNY Press, 2008), 109.

26. William Cullen Bryant, *The Embargo, or Sketches of the Times: A Satire, by a Youth of Thirteen* (1808; repr., with an intro. by Thomas O. Mabbott, Gainesville: University of Florida Press, 1955), quoted in Muller, *William Cullen Bryant*, 12.

27. Parke Godwin, *A Biography of William Cullen Bryant*, vol. 1 (New York: D. Appleton, 1883), 103, quoted in Muller, *William Cullen Bryant*, 18.

28. Tremaine McDowell, ed., *William Cullen Bryant* (New York: American Book Company, 1935), xxxiv. Many thanks are owed to my WPI colleague, the respected Poe scholar Kent Ljungquist, for sharing his thoughts and a trove of Bryant materials with me.

29. William Cullen Bryant, "A Yellow Violet" (1814), in *William Cullen Bryant: An American Voice*, ed. Frank Gado (White River Junction, VT: Antoca Press, 2006), 35.

30. William Cullen Bryant, "Green River" (1819), in Gado, *William Cullen Bryant*, 41–42.

31. William Cullen Bryant, "Thanatopsis" (1817, 1821), in Gado, *William Cullen Bryant*, 32.

32. Bryant, "Green River," 42.

33. Muller, *William Cullen Bryant*, 22–23.

34. William Cullen Bryant II and Thomas G. Voss, eds., *The Letters of William Cullen Bryant* vol. 1 (New York: Fordham University Press, 1975), 165.

35. Later in life Eaton published a series of newspaper articles on education. One recommended: "Girls should be taught like boys . . . 1. Mathematical Arts should never be neglected. 2. Physical structure of the earth, as deduced from geology. 3. The productions of the earth by which we are fed and clothed; . . . systems of botany and zoology. 4. That part of chemistry, which illustrates by experiments, the principles of domestic operations as they regard health and economy. 5. The philosophy of duty (or moral philosophy). 6. Thorough discipline in the art of communicating their thoughts with facility and grace." Amos Eaton, "Education, No. XII," *Troy Whig* (Apr. 21, 1835), quoted in McAllister, "Amos Eaton," 490.

36. Donald A. Ringe, "William Cullen Bryant and the Science of Geology," *American Literature* 26 (1955): 507–14.

37. Ibid., 507, 510.

38. William Cullen Bryant, "A Winter Piece" (1820), in Gado, *William Cullen Bryant*, 45.

39. William Cullen Bryant, "The Rivulet" (1823) in Gado, *William Cullen Bryant*, 54–56.

40. Muller, for example, discusses differences between the original poem and the 1821 revision as evidence of Bryant's emerging moral universe. See Muller, *William Cullen Bryant*, 23–24. Frank Gado offers his own mapping of the poems to Bryant's biography and stresses heavily how the grief process was manifested in Bryant's poem "Hymn to Death" (1820). See Gado, *William Cullen Bryant*, 11–15.

41. Bryant, "Thanatopsis," 32.

42. Eaton to Bryant, 21 June 1833 [Troy], in Bryant and Voss, *Letters*, 375–76, quoted in W. M. Smallwood, "Amos Eaton, Naturalist," *New York History* 18 (1937): 187; and also in McAllister, "Amos Eaton," 261. *Vâtes* is a Latin term meaning poet, but it also carries the connotation of seer or prophet.

43. Bryant to Eaton, 2 July 1833 [New York], in Bryant and Voss, *Letters*, 376; also quoted by McAllister, "Amos Eaton," 262.

44. Van Wyck Brooks, *The World of Washington Irving* (New York: E. P. Dutton, 1944), 195.

45. William Cullen Bryant, "The Prairie" (1833), in McDowell, *William Cullen Bryant*, 74–76. For the reference to contemporary scientific discourse, see, e.g., Caleb Atwater, "Description of the Antiquities Discovered in the State of Ohio and Other Western States," *Archaeologica Americana: Transactions and Collections of the American Antiquarian Society*, vol. 1 (Worcester, MA: William Manning, 1820). Ralph Miller's classic article analyzing Bryant's depiction of the mound builders has been widely cited over the years. "Nationalism in Bryant's 'The Prairies,'" *American Literature* 21 (1949): 227–32. Robert E. Bieder, *Science Encounters the Indian, 1820–1880* (Norman: University of Oklahoma Press, 1986), is still the best book on early Native American ethnology and linguistics, but we can look forward to fresh contributions by Sean Harvey, whose Ph.D. research broke new ground in these areas. "American Languages: Indians, Ethnology, and the Empire for Liberty," (Ph.D. diss., College of William and Mary, 2009), UMI publication 3392566.

46. William Cullen Bryant, "The Fountain" (1839), in McDowell, *William Cullen Bryant*, 95, quoted in Ringe, "William Cullen Bryant," 511.

47. William Cullen Bryant, "A Hymn of the Sea" (1842), in McDowell, *William Cullen Bryant*, 103, quoted and abridged further (than the intact excerpt I provide here) by Ringe, "William Cullen Bryant," 513.

48. Ringe, "William Cullen Bryant," 514.

49. Alan Taylor's masterful synthesis of social and literary history brings the details of William Cooper's fascinating career into sharp perspective, while examining how this history was translated into *The Pioneers* (1823), the first of James Cooper's Leather-Stocking novels. *William Cooper's Town: Power and Persuasion on the Frontier of the Early American Republic* (New York: Vintage, 1995), 113.

50. Wayne Franklin, *James Fenimore Cooper: The Early Years* (New Haven: Yale University Press, 2007), 40–41.

51. Franklin, *James Fenimore Cooper*, 51.

52. Taylor, *William Cooper's Town*, 105.

53. George P. Fisher, *Life of Benjamin Silliman*, vol. 1 (New York: Charles Scribner, 1866), 127.

54. Wayne Franklin analyzes the circumstances surrounding Cooper's "gunpowder plot" in far greater detail. See Franklin, *James Fenimore Cooper*, 54–58.

55. Franklin points out that Cooper was not as awed by the falls at Niagara on his first trip as he was on subsequent visits, after his first extended residence in Europe had taught him to regard landscapes through the Romantic aesthetic lens. *James Fenimore Cooper*, 121.

56. Nelson Adkins wrote the classic summary of Cooper's role in founding and

sustaining the Bread and Cheese Club. "James Fenimore Cooper and the Bread and Cheese Club," *Modern Language Notes* 47 (1932): 71–79.

57. John Finch, *Travels in the United States of America and Canada, containing some accounts of their scientific institutions, and a few notices of the geology and mineralogy of those countries, to which is added An Essay on the Natural Boundaries of Empires* (London: Longman, Rees, Orme, Brown, Green, and Longman, 1833), 29.

58. For more on British Romantic figures and the study of geology, see Noah Heringman, *Romantic Rocks: Aesthetic Geology* (Ithaca, NY: Cornell University Press, 2004), and Dennis R. Dean, *Romantic Landscapes: Geology and Its Cultural Influence in Britain, 1765–1835* (Evanston, IL: Scholars' Facsimiles and Reprints, 2007).

59. Ethel McAllister alludes on three occasions to a Professor William Cooper, and although this New York conchologist was an occasional attendee of the Bread and Cheese Club, he was in all likelihood not even a cousin to James Fenimore Cooper. McAllister, "Amos Eaton," 266, 295, 325; Franklin, *James Fenimore Cooper*, 317.

60. According to views first put forward by Orm Överland in *The Making and Meaning of an American Classic: James Fenimore Cooper's The Prairie* (New York: Humanities Press, 1973), Cooper relied upon expedition naturalist Edwin James's published account of the 1819 Long Expedition for his factual knowledge of the trans-Mississippi West. See Edwin James, *Account of an Expedition from Pittsburgh to the Rocky Mountains, Under the Command of Major Stephen H. Long* (Philadelphia: H. C. Carey and I. Lea, 1823; repr., Barre, MA: Imprint Society, 1972). Lance Schachterle inclines, however, toward James P. Elliott, who argues that Cooper went far beyond reproducing chunks of this primary source when imagining scenery and composing landscape descriptions for *The Prairie*. My claim here is simpler and yet a key point possibly underappreciated by these literary scholars. Cooper was an educated reader of scientific texts. It is noteworthy that he was in no way embarrassed to disclose what he learned about geology to the readers of his fictional works. See Lance Schachterle, "On *The Prairie*," in *Leather-Stocking Redux; Or, Old Tales, New Essays*, ed. Jeffrey Walker (New York: AMS Press, 2011), 124–49, see esp. 126.

61. James Fenimore Cooper, *The Prairie* (1827; rev. with a new intro., London: Richard Bentley, 1836; repr., New York: Co-operative Publication Society), 37.

62. Cooper, *The Prairie*, 223.

63. Ibid., 194.

64. Franklin, *James Fenimore Cooper*, 51.

65. The scholarly literature on Rafinesque, beginning with his prolific self-promotion and the caustic reactions that it provoked, is voluminous and extremely colorful. Historian of biology Jim Endersby published a balanced and useful reconsideration of the factors that account for all the sound and fury that have surrounded Rafinesque almost since the moment he began to conduct field research on American natural history. "'The Vagaries of a Rafinesque': Imagining and Classifying American Nature," *Studies in History and Philosophy of Biological and Biomedical Sciences* 40 (2009): 168–78. A definitive biography has also recently been published by Rafinesque scholar Charles Boewe; see *The Life of C. S. Rafinesque, A Man of Uncommon Zeal* (Philadelphia: American Philosophical Society, 2011).

66. This identification between Nuttall and Obed Bat, though an authentic original reaction facilitated by the juxtaposition of this chapter's examination of Irving, Bryant,

and Cooper's separate renditions of exploration of the prairie, is not mine alone. Cooper scholar Robert Madison broached the same claim in 1993, and he has since tried to document occasions in New York where Cooper may have been able to encounter Nuttall in person as early as 1822. "Cooper and Nuttall: The Course of Empire," in Steven Harthorn and Shalicia Wilson, eds., *James Fenimore Cooper Society Miscellaneous Papers* 26 (2009): 9–10, available at http://external.oneonta.edu/cooper/articles/ala/2009ala-madison.html.

67. Lance Schachterle, "Cooper, Style, and *The Bravo*," Oneonta Cooper Conference Keynote Address (Aug. 2009), in *James Fenimore Cooper: His Country and His Art, Papers from the 2009 Cooper Conference*, no. 17 (Oneonta: SUNY College at Oneonta, 2011), 46–64, see esp. 54.

68. Susan Fenimore Cooper, *Pages and Pictures from the Writings of James Fenimore Cooper* (New York: W. A. Townsend and Co., 1861; repr., New York: Castle Books, 1980), 164–65.

69. Cooper, introduction to *The Prairie*, 5.

70. Historian of science Bert Hansen outlined the relevant timeline for British geological debates about floating ice in his classic historical review article. "The Early History of Glacial Theory in British Geology," *Journal of Glaciology* 9 (1970): 137.

71. Adams cites James Fenimore Cooper, *The Crater; or Vulcan's Peak, a Tale of the Pacific* (1847; repr., New York: W. A. Townsend and Co., 1861), 176. Charles H. Adams, "Uniformity and Progress: The Natural History of *The Crater*," in *James Fenimore Cooper: New Historical and Literary Contexts*, ed. W. M. Verhoeven (Amsterdam: Rodopi, 1993), 213. Incidentally, Harry Hayden Clark had similarly identified *The Crater* as the pinnacle of Cooper's incorporation of geology as a literary device: "He had an allegorical purpose to serve, but the geological processes–the rise and sinking of the island–serve to keynote the twin theses that God governs all and that religion must be prior to politics. For his descriptions of the cataclysmic actions of nature, Cooper looked to what scientific inquiry, particularly the researches of Sir Charles Lyell, had provided, with the result that his story had a probity it otherwise might have lacked." "Fenimore Cooper and Science, Part II," *Transactions of the Wisconsin Academy of Sciences, Arts and Letters* 49 (1960): 272–73.

72. James Fenimore Cooper, *Notions of the Americans: Picked up by a Travelling Bachelor*, vol. 1 (1828; repr., Philadelphia: Carey, Lea, and Blanchard, 1835), 204.

73. Ibid., 244.

74. For example, "Americans lacked what Europeans had–the sustaining sense of history, and a decorum bred by tradition." Lewis Leary, *Soundings: Some Early American Writers* (Athens: University of Georgia Press, 1975), 315.

75. Hedges, *Washington Irving*, 115.

76. Brooks, *World*, 199.

77. Cooper, *Notions of the Americans: Picked up by a Travelling Bachelor*, vol. 2 (London: Henry Colburn, 1828), 142.

78. Norman Foerster, *Nature in American Literature: Studies in the Modern View of Nature* (New York: Russell and Russell, 1958), 5–6.

79. Studies of Irving's interest in the American Indian usually feature his essay "Traits of Indian Character" (1819) and otherwise touch upon relevant portions of the nonfiction works, *Tour of the Prairies, Astoria, and The Adventures of Captain Bonneville*. A nice

overview and response to this scholarship can be found in Daniel F. Littlefield Jr., "Washington Irving and the American Indian," *American Indian Quarterly* 5 (1979): 135–54. Bryant's views on the politics and morality of Indian removal have been discussed in passing in this book, but a more focused discussion can be found in Joel Pace, "William Wordsworth, William Cullen Bryant, and the Poetics of American Indian Removal," in *Native Americans and Anglo-American Culture, 1750–1850; The Indian Atlantic*, ed. Tim Fulford and Kevin Hutchings (Cambridge: Cambridge University Press, 2009), 197–216. The scope of scholarship on Cooper's treatment of Indians as a fruitful subject is too voluminous to begin to list in any systematic manner. Two sources that present critically valuable arguments for historians include Roy Harvey Pearce, *Savagism and Civilization: A Study of the American Indian Mind* (Baltimore: Johns Hopkins University Press, 1965), 196–236, and Susan Scheckel, *The Insistence of the Indian: Race and Nationalism in Nineteenth Century American Culture* (Princeton: Princeton University Press, 1998), 15–40.

80. This confirmation comes from Foerster: "Aside from Irving, our first literature of considerable absolute value, that of Bryant and Cooper, is therefore in a very real sense the expression of the first distinction of America, her physical self." Foerster, *Nature*, 3–4.

CHAPTER 8: Kindred Spirits

Epigraph. William Cullen Bryant, *On the Life of Thomas Cole* (a funeral oration delivered at the National Academy of Design, New York, 4 May 1848), www.catskillarchive.com/cole/wcb.htm.

1. A recent paper by art historian Andrew Graciano, though it focuses on British rather than American subjects, provides an exemplary model of analysis. "'The Book of Nature is Open to All Men': Geology, Mining, and History in Joseph Wright's Derbyshire Landscapes," *Huntington Library Quarterly* 68 (2005): 583–600.

2. One of the first clues that led me to examine the broader cultural impact of geological ideas about time and scientific observations of rocks came from art historian Barbara Novak. *Nature and Culture: American Landscape and Painting, 1825–1875* (New York: Oxford University Press, 1980), 58.

3. Rebecca Bedell, *The Anatomy of Nature: Geology and American Landscape Painting, 1825–1875* (Princeton: Princeton University Press, 2002).

4. Kenneth Haltman, "The Poetics of Geologic Reverie: Figures of Source and Origin in Samuel Seymour's Landscapes of the Rocky Mountains," *Huntington Library Quarterly* 59 (1996): 303–47.

5. David S. Reynolds distinguishes the work of fellow Hudson River landscape artists Asher Durand and Thomas Doughty from Cole's precisely in terms of their more single-minded embrace of the near-photographic attempt to capture the minutiae of nature. *Walt Whitman's America: A Cultural Biography* (New York: Alfred A. Knopf, 1995), 296.

6. Originally commissioned by New York dry-goods merchant and art collector Jonathan Sturgis, Asher Durand's *Kindred Spirits* (1849) was held in trust for a century by the New York Public Library as a private donation by William Cullen Bryant's daughter Julia, but now the painting belongs to the Walton Family Foundation as a result of a silent bid auction. See U.S. National Gallery of Art, "Asher Durand's *Kindred Spirit*," exhibition notes, July 1, 2005–Mar. 15, 2007, www.nga.gov/exhibitions/durandinfo.shtm.

7. *Lake With Dead Trees (Catskill)* (1825) now belongs to the Allen Memorial Art Museum permanent collection at Oberlin College. Image and background information

are available at www.oberlin.edu/amam/Cole.htm (accessed 22 Aug. 2011). For a summary of Cole's professional debut and hearty reception by Dunlap et al., see Ellwood C. Parry III, *The Art of Thomas Cole: Ambition and Imagination* (Newark: University of Delaware Press, 1988), 23–27.

8. Bryant's funeral oration contains a famously nasty paragraph exposing how Featherstonhaugh financially abused Cole:

> It was the fate of Cole, at this period of his life, to meet with a patron. When his pictures first attracted the public attention, as I have already related, a dashing Englishman, since known as the author of a wretched book about the United States, who had married the heiress of an opulent American family, professed to take a warm interest in the young painter, and charged himself with the task of advancing his fortunes. He invited him to pass the winter at his house, on his estate in the country, and engaged him to paint a number of landscapes, for which he was to pay him twenty or thirty dollars each, a trifling compensation for such works as Cole could even then produce, but which I have no doubt, seemed to him at that time munificent. It would hardly become the place or the occasion were I to relate the particulars of the treatment which the artist received from his patron, the miserable and cheerless apartments he assigned him, the supercilious manner by which he endeavored to drive him from his table to take his meals with the children of the family, and the general disrespect of his demeanor.

Bryant, *On the Life of Thomas Cole*. Parry notes that Featherstonhaugh intended to use Cole's paintings of Featherston Park to entice foreign investors to fund development of his land holdings in Duanesburg. *The Art*, 34. For Featherstonhaugh's side of the story, one can find a sympathetic treatment in Edmund Berkeley and Dorothy Smith Berkeley, *George William Featherstonhaugh: The First U.S. Government Geologist* (Tuscaloosa: University of Alabama Press, 1988), 38–39.

9. Parry indicates that Cole's introduction to the Patroon took place in a letter dated 8 Nov. 1826. *The Art*, 40. Christine Robinson adds that although the relationship "was never fully developed," Cole did execute two landscapes commissioned by Van Rensselaer: *View on Lake Winnepesaukee* (1828) and *View Near Catskill* (1829). Robinson, "Thomas Cole: Drawn to Nature," in *Thomas Cole: Drawn to Nature* (Albany: Albany Institute of History and Art, 1993), 63. The Patroon confirmed after the Winnepesaukee painting had been safely delivered to Albany: "Mrs. R. appears well pleased with it." Van Rensselaer to Cole, 5 July 1828, quoted in Parry, *The Art*, 80.

10. Robinson, "Thomas Cole," 65.

11. Novak, *Nature*, 57.

12. These artifacts, which attest to the care and curiosity of the artist as a geological specimen hunter, are now housed at the Thomas Cole Historic Site in Catskill, New York. "Thomas Cole's New York Fossil and Rock Collection, 1830–45," accessed 3 Dec. 2012, www.explorethomascole.org/scrapbook/items/221.

13. Barbara Novak, *American Painting of the Nineteenth Century: Realism, Idealism, and the American Experience* (New York: Harper and Row, 1979), 77–78.

14. Ellwood Parry, it should be noted, disagrees with the extremity of Novak's characterization of Cole's behavior, saying it "hardly fits the facts." *The Art*, 340.

15. Bedell, *Anatomy*, 5.

16. Note that the French classicist painter Nicholas Poussin's mid-seventeenth-century masterpiece *Winter (The Flood)* anticipated Romanticism and its "horrific sublime" treatment of the apocalyptic power of nature by a good century. At the beginning of the nineteenth century, a steady parade of landscape artists took their turn at rendering Romantic visions of the Deluge: J. M. W. Turner's *The Deluge* (1804–5) and Théodore Géricault's *The Deluge* (1818) both predated Martin's highly regarded attempt. For more on Poussin's influential legacy, see Richard Verdi, "Poussin's 'Deluge': The Aftermath," *Burlington Magazine* 123 (1981): 389–401.

17. Review of *The Deluge*, by John Martin, *The Critic: A Weekly Review of Literature, Fine Arts, and Drama* (3 Jan. 1829), excerpted from *London New Monthly Magazine* (Oct. 1828) quoted in Parry, *The Art*, 87–88.

18. Parry's analysis of *The Subsiding of the Waters of the Deluge* assigned to Cole a sympathy with both Neptunism and catastrophism, but he made these attributions in the absence of a more historically informed understanding of how loosely defined these ideologies were at the time. Parry likely had not seen the painting in person, since it was only made available again to public view in 1983. Parry, *The Art*, 60.

19. "Review Exhibition" notice of the National Academy of Design in the *New-York Mirror*, 7 May 1831: 350, quoted in the caption to plate 6 in Parry, *The Art*, 128–29.

20. Régis Gignoux's *Mammoth Cave, Kentucky* (1843) belongs to the New-York Historical Society and was included among the Hudson River landscape artists exhibit at the Peabody Essex Museum in the summer of 2011 at Salem, Massachusetts. For an online image and some commentary on the restoration of the painting, see http://antiquesand thearts.com/TT0–07–25–2000–15–05–19 (accessed 31 Aug. 2011).

21. William Cullen Bryant, "To Cole, the American Painter, Departing For Europe" (1829), reprinted in John W. McCoubrey, ed., *American Art, 1700–1960: Sources and Documents* (Englewood Cliffs, NJ: Prentice-Hall, 1965), 96.

22. Novak, *Nature*, 56; Bedell, *Anatomy*, 18. Of Comstock's views, Benjamin Silliman felt compelled to temporize in his review: "Geology contradicts almost nothing in the sacred history;–all that it requires is an extension of time; whereas the modern astronomy is in almost exact opposition to the literal sense of the language of the bible; still no one now dreams of any real inconsistency between them." Review of *Outlines of Geology*, by J. L. Comstock, *AJS* 26 (1834): 212–13.

23. Thomas Cole's *Distant View of Niagara Falls* (1830) belongs to the Art Institute of Chicago, image available at www.explorethomascole.org/gallery/items/43 (accessed 31 Aug. 2011).

24. Bostonian Louisa Davis Minot's only two known paintings, both of Niagara Falls, derived from a trip she made to the popular location. Her account of this trip was published as "Sketches of Scenery on Niagara River," *North American Review* 2 (1816): 320–29. Her *Niagara Falls* (1818), which belongs to the New-York Historical Society, was included among the Hudson River landscape artists exhibit at the Peabody Essex Museum. Contrary to my impression, the exhibit curator's notes assert that Minot "exploits a sense of the sublime in these paintings, a sense of awe and even fear at the overwhelming power of nature on a grand scale." See Linda S. Ferber, *The Hudson River School: Nature and the American Vision* (New York: New-York Historical Society, 2009), 123. Image available at www.pegasusnews.com/news/2011/mar/01/hudson-river-school-art -review-amon-carter/ (accessed 1 Sept. 2011).

25. Alvan Fisher's *A General View of the Falls of Niagara* (1820) belongs to the National Museum of American Art collection.

26. Alvan Fisher's *The Great Horseshoe Fall, Niagara* (1820) belongs to the Smithsonian American Art Museum collection.

27. Alvan Fisher's *Niagara; The American Falls* (1821) belongs to the New-York Historical Society and was included among the Hudson River landscape artists exhibit at the Peabody Essex Museum.

28. Parry, *The Art*, 145. I had the opportunity to examine the *Course of Empire* as part of the Hudson River landscape artists exhibit at the Peabody Essex Museum in Salem, Massachusetts. This experience really confirms my appreciation for Parry's analysis. Not having thought about what Parry wrote in many years, I recorded my independent reactions to what I was seeing in Cole's paintings: "All five scenes are dramatically different in their subject matter and featured content, but Cole's prominent use of a rocking stone perched atop a rocky crag in every canvas signifies not only that these situations all share the same physical space but also that the passage of time, otherwise so apparent and dramatic in terms of the transformative effects of human activity, nevertheless has been telescoped into an inconsequential interval of time as the planet experiences it (geologically)." David Spanagel, journal, 23 Aug. 2011.

29. Thomas Cole, "Essay on American Scenery," *American Monthly Magazine* 1 (1836): 1–12; reprinted in McCoubrey, *American Art*, 98–110, quotation from 101.

30. William Cullen Bryant to William Ware, 11 Oct. 1834 [Florence], in vol. 1 of *The Letters of William Cullen Bryant*, ed. William Cullen Bryant II and Thomas G. Voss (New York: Fordham University Press, 1975), 426–27.

31. Thomas Cole's *The Vale and Temple of Segesta, Sicily* (1844) was also on display at the Hudson River landscape artists exhibit at the Peabody Essex Museum in the summer of 2011 at Salem, Massachusetts. In 1842 Cole prepared a pen and brown ink over graphite pencil study in advance of composing the finished painting. An original of the study now belongs to the Detroit Institute of Arts (accession number 39.421.A); image available at www.dia.org/user_area/comping/39.421.A-D1.jpg (accessed 31 Aug. 2011).

32. Thomas Cole to George W. Greene (date unknown but sent after Cole's return to New York in 1842), quoted by Van Wyck Brooks, *The World of Washington Irving* (New York: E. P. Dutton, 1944), 362 n.

33. Thomas Cole's *Catskill Mountain House: The Four Elements* (1843–44, belongs to a private collection. Image available at http://home.comcast.net/~e.hartouni/img/ Cole_Thomas_Catskill_Mountain_House_The_Four_Elements_1843–44.jpg (accessed 7 Sept. 2011). Thomas Cole's *Catskill Creek, New York* (1845) belongs to the New-York Historical Society and was included among the Hudson River landscape artists exhibit at the Peabody Essex Museum. Detailed image available at http://3.bp.blogspot.com/ -iFCT1NoEgis/Ti87AyqAu4I/AAAAAAAAJ-I/TPnEzAAQnXo/s1600/o8.jpg (last accessed 7 Sept. 2011).

34. Parry spells out the terms of Cole's tutelage. For an annual tuition of $300, plus $3 a week for room and board, Cole took Church on as a student in 1844 and added Benjamin Conkey as a second pupil in the fall of 1845. Ellwood C. Parry III, "Thomas Cole's 'The Hunter's Return'" *American Art Journal* 17 (1985): 15.

35. Bedell explores the twin lines of this parting observation to a satisfying degree in her chapters "Asher Durand and the Therapeutic Landscape" and "Frederic Church

and the Educational Enterprise." See Bedell, *Anatomy*, 47–83. For more on the geological vision Church explored, especially in his paintings of the Andes, see the classic essay by Stephen Jay Gould, "Church, Humboldt, and Darwin: The Tension and Harmony of Art and Science," in *Frederic Edwin Church*, ed. Franklin Kelly (Washington: Smithsonian Institution Press, 1989), 184–97.

CHAPTER 9:   Rocks, Reverence, and Religion

Epigraph. D. C. Smith to Amos Eaton, 10 Aug., 1835 [Schenectady], folder 4, box 1, Amos Eaton Papers.

1. I use the term *trading zone* here less in the sense of its Galisonian appropriation to describe encounters of twentieth-century physicists and more according to its basic anthropological roots. In other words, despite their (emerging) differences in language and belief systems, students of natural history and champions of a literal reading of the Christian scriptures were engaging in public encounters that highlighted both the potential for conflict and the participants' evolving understandings of their own commitments to either a harmonic compromise or triumphant outcome. For a retrospective analysis of Galison's use of the term, see Peter Galison, "Trading With the Enemy," in *Trading Zones and Interactional Expertise: Creating New Kinds of Collaboration*, ed. Michael E. Gorman (Cambridge, MA: MIT Press, 2010), 25–52.

2. Benjamin Silliman, "Supplement by the Editor," appended to Robert Bakewell, *An Introduction to Geology: Intended to Convey a Practical Knowledge of the Science, and comprising the most important recent discoveries; with explanations of the facts and phenomena which serve to confirm or invalidate various geological theories* (New Haven: Hezekiah Howe, 1833), 389–91.

3. Walter H. Conser Jr., *God and the Natural World: Religion and Science in Antebellum America* (Columbia: University of South Carolina Press, 1993), 16.

4. Landmark books on this topic include Whitney R. Cross, *The Burned-Over District: The Social and Intellectual History of Enthusiastic Religion in Western New York, 1800–1850* (Ithaca, NY: Cornell University Press, 1950); Paul E. Johnson, *A Shopkeeper's Millennium: Society and Revivals in Rochester, New York, 1815–1837* (New York: Hill and Wang, 1979); Michael Barkun, *Crucible of the Millennium: Burned-Over District of New York in the 1840s* (Syracuse, NY: Syracuse University Press, 1986); and John L. Brooke, *The Refiner's Fire: The Making of Mormon Cosmology, 1644–1844* (Cambridge: Cambridge University Press, 1996).

5. Barkun, *Crucible*, 23.

6. Amos Eaton, journal, vol. 1 (Williams College), 21, 23, and 36, folder 11, box 6, Eaton Papers. Note that Eaton's compulsory revival attendance predates by almost a decade the leading edge of the Second Great Awakening, according to Barkun's reckoning in *Crucible*, 23.

7. George Johnson, diary, 15 May 1836, folder 1, box 5, Eaton Papers. Johnson was a student at the Rensselaer Institute. Because he eventually married one of Eaton's daughters, Johnson's diary for 1836–37 is among the Eaton family's papers.

8. George Johnson, diary, 22 May 1836, folder 1, box 5, Eaton Papers.

9. See Edward Hitchcock, "Report on the Geology of Massachusetts," *AJS* 21 (1832): 1–70.

10. Hitchcock's paper was read before the American Geological Society on 11 Sept.

1822, and an expanded version of his report was subsequently published in installments in *Silliman's Journal*. Edward Hitchcock, "A Sketch of the Geology, Mineralogy, and Scenery of the Regions contiguous to the River Connecticut," *AJS* 6 (1823): 1–86, 201–36, and *AJS* 7 (1824): 1–30, quotation from page 17 of the final installment.

11. John Phillips, *Illustrations of the Geology of Yorkshire* (1829) was endorsed by Benjamin Silliman in "Principles of Geology," *AJS* 21 (1832): 1–26, quotations from pages 14–16.

12. See Benjamin Silliman, "Outline of a Course of Geological Lectures, Given at Yale College," appended to Robert Bakewell, *An Introduction to Geology: Comprising the Elements of the Science in its Present Advanced State* (New Haven: Hezekiah Howe, 1829); and Rodney Lee Stiling, "The Diminishing Deluge: Noah's Flood in Nineteenth-century American Thought" (Ph.D. diss., University of Wisconsin-Madison, 1991), 93.

13. Amos Eaton, *Geological Journal F*, 30 June 1829 [Nevisink, New Jersey], unpaginated, folder 4, box 2, Eaton Papers.

14. Ibid., 25 June 1829 [New York].

15. For more on Hitchcock and the flightless birds theory, see Nancy Pick and Frank Ward, *Curious Footprints: Professor Hitchcock's Dinosaur Tracks and Other Natural History Treasures at Amherst College* (Amherst, MA: Amherst College Press, 2006).

16. Edward Hitchcock, *The Religion of Geology and Connected Sciences* (Boston: Phillips, Sampson and Company, 1854), 23–24.

17. Key works that provided a scientific underpinning to this reactionary Christian intellectual response include the Seventh Day Adventist scholar George McCready Price's foundational textbook *The New Geology* (Nampa, ID: Pacific Press Publishing Association, 1923), and fundamentalist Christian theologian John C. Whitcomb Jr. and hydraulic engineer Henry M. Morris's jointly written update of McCready's argument, *The Genesis Flood: The Biblical Record and Its Scientific Implications* (Phillipsburg, NJ: P and R Publishing, 1961). For an excellent and balanced history of science analysis of twentieth-century developments along these lines, see Ronald L. Numbers, *The Creationists: The Evolution of Scientific Creationism* (Berkeley: University of California Press, 1993).

18. Hitchcock, *Religion*, 122.

19. Thomas Church Brownell, *Soldier and Servant: Autobiography of Thomas Church Brownell, Third Bishop of Connecticut* (1858; repr., with an intro. by William A. Beardsley, Hartford, CT: Church Mission Publishing Company, 1924), 10.

20. Biographical details taken from William Agur Beardsley, "Thomas Church Brownell," *Dictionary of American Biography* 3 (1929): 171–72.

21. Thomas Church Brownell, "Missionary Journal," reprinted in *Protestant Episcopal Church History Journal* 7 (1938): 303–22, cited by Robert Rogers Hubach, *Early Midwestern Travel Narratives: An Annotated Bibliography, 1634–1850* (Detroit: Wayne State University Press, 1961), 67.

22. D. C. Smith to Amos Eaton, 10 Aug. 1835, folder 4, box 1, Eaton Papers.

23. Ibid.

24. The idea of a war between science and religion was quite alien to Eaton and Hitchcock's generation. It gained steam with the publication of J. W. Draper's *History of the Conflict Between Religion and Science* (New York: Appleton, 1874) and Andrew Dickson White's *A History of the Warfare of Science With Theology in Christendom*,

2 vols. (New York: Appleton, 1896). These two polemics by then-prominent academic writers set the tone for much of the subsequent historical scholarship on the relations between geology and religion until Charles Coulston Gillispie's *Genesis and Geology: A Study in the Relations of Scientific Thought, Natural Theology and Social Opinion in Great Britain, 1790–1850* finally came along (1951; repr., Cambridge: Harvard University Press, 1996). For a thorough historical analysis of the warfare metaphor, see James R. Moore, *The Post-Darwinian Controversies: A Study of the Protestant Struggle to Come to Terms With Darwin in Great Britain and America, 1870–1900* (Cambridge, MA: Harvard University Press, 1979), 19–124.

25. Conser broadly examines the influence of contemporary German, English, and Scottish theologians on American Protestant culture. Despite his primary focus on the beliefs of religious leaders in the southern United States, he does pay attention to the philosophical and theological influence of contemporary geological ideas, including those advanced by Hitchcock and Silliman. (Notably, he does not mention Amos Eaton.). Conser, *God*, 8–36.

26. Stiling, "Diminishing Deluge," 11.

27. Ibid., 12–13.

28. Amos Eaton, *Eaton's Geological Note Book, for the Troy Class of 1841*, 5, 12, folder 8, box 1, Eaton Papers.

29. Edward Hitchcock to Benjamin Silliman, 12 Mar. 1837, quoted in Gloria Robinson, "Edward Hitchcock," in *Benjamin Silliman and His Circle: Studies on the Influence of Benjamin Silliman on Science in America*, ed. Leonard G. Wilson (New York: Science History Publications, 1979), 67.

30. Amos Eaton, "Geological Nomenclature, Classes of Rocks, &c.," *AJS* 14 (1828): 359–60.

31. J. G. Ambler, M.D., "The Class of Thirty-Three," *Proceedings of the Semi-Centennial Celebration of the Rensselaer Polytecnic Institute* (Troy, NY: William H. Young, 1875), 108.

32. Martin I. Townsend, *Proceedings of the Semi-Centennial Celebration*, 40.

33. Ethel McAllister supported this conclusion: "That he was deeply religious in the true sense of the word cannot be doubted. He was constantly trying to reconcile . . . himself and his scientific, inquiring mind to some of the dogmas of the Christian faith." "Amos Eaton: Scientist and Educator" (Ph.D. diss., University of Pennsylvania, 1941), 62–63.

CONCLUSION: Echoes of New York's Embrace of Geological Investigation

1. A recently published exception underscores the general rarity of such works. See Andrew J. Lewis, *A Democracy of Facts: Natural History in the Early Republic* (Philadelphia: University of Pennsylvania Press, 2011). Lewis contends that Americans were fascinated by matters of natural history during the early national period, but his book argues that the process of scientific professionalization and the cultivation of respectability for the practices of natural history were necessarily tied to the employment of scientists in service to the bureaucracies of state governments. As I have tried to show, the case of natural history in New York is not quite so simple or straightforward. Notably, Lewis confines the entire story of Amos Eaton's geological, institutional, and practical innovations to a single paragraph. *A Democracy*, 139–40.

2. Robert Siegfried to David Spanagel, 10 Nov. 1995, private correspondence.

3. Timothy Dwight, *Travels in New-England and New-York,* vol. 4 (New Haven: Timothy Dwight, 1822), 220. See also C. F. Volney, *A View of the Soil and Climate of the United States of America* (London: J. Johnson, 1804; repr., with a new intro. by George W. White in New York: Hafner, 1968).

4. Dwight, *Travels,* 4:235. See also François A. F. de la Rochefoucauld-Liancourt, *Voyages dans les Etats Unis d'Amerique fait en 1795, 1796 et 1797,* 8 vols. (Paris, 1799).

5. Edward Hitchcock completed the Massachusetts survey report in 1830. Edward Allen, Lewis Beck, George Boyd, Ebenezer Emmons, James Hall, and Eben Horsford all worked for the New York Natural History Survey when it was launched in 1836, with Hall eventually serving as state geologist for New York, Iowa, and Wisconsin, while Emmons went on to become North Carolina's state geologist. James Curtis Booth assisted Henry Darwin Rogers on the Pennsylvania survey before Booth became Delaware's state geologist. Caleb Briggs worked with W. W. Mather on the Ohio survey and with William Barton Rogers on Virginia's survey. Ezra Carr eventually joined Hall as Wisconsin's state geologist. George Cook made his career as New Jersey's state geologist. Douglass Houghton founded and led the Michigan geological survey until his tragic death. Michael Tuomey served as state geologist in Alabama and South Carolina. Thanks to the Rensselaer Polytechnic Institute Department of Earth and Environmental Sciences for compiling this list of alumni achievements. See http://ees2.geo.rpi.edu/History/state_geo.html (accessed Oct. 2011).

Essay on Sources

The initial inspiration for my study of geological thinking in early nineteenth-century New York State came out of the broader investigation into the history of the American earth sciences that I launched in the early 1990s. My 1996 Ph.D. dissertation, "Chronicles of a Land Etched by God, Water, Fire, Time, and Ice," constituted my first attempt to frame into some kind of narrative structure my ideas about the cultural history of American geology. My deep regard for the integrity of primary source materials and my astonishment at both the ingenuity of and the multivalent influences on the early nineteenth-century scientific mind were born when I broached my research topic idea to Lisa Herschbach, then a graduate student colleague in the history of science department at Harvard University. I told Lisa that I wanted to learn how and why practitioners of the natural sciences became prominent and respectable holders of social authority in a culture like that of the United States, where dogmatic religious belief systems and a cult of uneducated democratic self-reliance would seem to be major obstacles to such a development.

Lisa helpfully suggested that I consult with historian Mott Greene, the celebrated author of monumental scholarly works on the history of European geology and a professor at the University of Puget Sound, whom she knew personally to be a wise and sensitive human being. Greene's advice was simply put: If I wanted to learn anything about the beginnings of American science, I should start by reading the issues of Benjamin Silliman's *American Journal of Science* cover to cover from 1818 forward. So I did that. Virtually becoming an appendage of the microfilm reader at Harvard's Kummel Geology Library for the next sixteen months, I scoured the seemingly endless series of negative page images (white text on a black screen), transcribing every provocative observation and theory that I could find about natural phenomena and piecing together the chains of new ideas that were stimulated by exploring the interior of North America in the 1810s, 1820s, and 1830s. Ultimately, this exhaustive preliminary research assignment yielded over 150 pages of handwritten notes of my own covering articles published during the first twenty-one years or so of that periodical's existence. The laborious apprenticeship that Greene had casually recommended from afar provided two key things: it enabled me to become deeply familiar with dozens of key figures, and it equipped me with an indelible sense of both the boundless fecundity and the total lack of discipline that characterized early nineteenth-century American scientific investigations.

A wide array of possible research foci now lay before me, but I chose to delve most deeply into the history of geology because the issue I had initially cared most about (the potentially destabilizing nature of scientific knowledge) looked most likely to be exposed. Throughout the primary source materials I had examined, questions regarding the Earth's history and the processes of geophysical dynamics sometimes provoked a wider sense of danger and disharmony between the natural sciences and other prevailing forms of cultural authority. To situate my provocative American episodes within their larger intellectual and political contexts, I needed to obtain a far more solid understanding of the global history of geology. I scoured the field for foundational works of scholarship and synthesis and was gradually brought up to speed by reading the three volumes of Charles Lyell, *Principles of Geology; Being an Attempt to explain the Former Changes of the Earth's Surface, by Reference to Causes Now in Operation* (London: John Murray, 1830–1833; ed. and with a new intro. by Martin J. S. Rudwick, Chicago: University of Chicago Press, 1990); John Imbrie and Katherine Palmer Imbrie, *Ice Ages* (Cambridge, MA: Harvard University Press, 1979); Nicolaas A. Rupke, *The Great Chain of History: William Buckland and the English School of Geology, 1814–1849* (Oxford: Clarendon Press, 1983); Paolo Rossi, *The Dark Abyss of Time*, trans. from the Italian by Lydia G. Cochrane (Chicago: University of Chicago Press, 1984); James A. Secord, *Controversy in Victorian Geology* (Princeton: Princeton University Press, 1986); Stephen Jay Gould, *Time's Arrow Time's Cycle* (Cambridge, MA: Harvard University Press, 1987); Rachel Laudan, *From Mineralogy to Geology* (Chicago: University of Chicago Press, 1987); Gabriel Gohau, *A History of Geology*, rev. and trans. from the French by Albert V. Carozzi and Marguerite Carozzi (New Brunswick, NJ: Rutgers University Press, 1990); David R. Oldroyd, *The Highlands Controversy* (Chicago: University of Chicago Press, 1990); and Martin J. S. Rudwick, *Scenes from Deep Time* (Chicago: University of Chicago Press, 1992). Mott Greene's own excellent survey work *Geology in the Nineteenth Century* (Ithaca, NY: Cornell University Press, 1984) stood out from this pack, opening my eyes to the fact that not all important developments in the history of geology were accomplished by native English speakers. Were I to undertake the same task of getting oriented to literature in the history of geology today, I would add Naomi Oreskes, *The Rejection of Continental Drift* (New York: Oxford University Press, 1999), and Pascal Richet, *A Natural History of Time*, trans. from the French by John Venerella (Chicago: University of Chicago Press, 2007), to my list of obligatory background readings.

Back in the early 1990s, writings on the history of geology outside of Europe were scanty, even compared with those concerning non-English European traditions. Furthermore, what little was written about American topics tilted heavily toward an antique mode of celebration (chronicles by retired scientists rather than analyses by trained historians of science). One valuable collection that introduced me to important colleagues among those science historians, however, was the conference volume *Two Hundred Years of Geology in America*, ed. Cecil J. Schneer (Hanover, NH: University Press of New England, 1979). Specific contributors to this volume, whose interests would directly feed my own emerging research agenda, included Michele L. Aldrich, Harold L. Burstyn, Patsy A. Gerstner, Ellwood C. Parry III, Stephen J. Pyne, and Kenneth L. Taylor. Once I located and devoured Michele Aldrich's 1974 University of Texas-Austin history Ph.D. dissertation (on microfilm), a tremendous piece of work that was long delayed in its publication as a book (see below), I knew that I had something really interesting and fruitful to pursue in the question "What happened to engage New York's interest in its geology, prior to the creation of a publicly-funded professional natural history survey effort?"

PRIMARY SOURCES

The backbone of this study relied on archival research conducted primarily at the New York State Library in Albany. Over the course of multiple visits, I was able to work my way systematically through all of the boxes and folders stuffed with journals, correspondence, clippings, and miscellaneous items that related directly to Amos Eaton's career. This research included the Amos Eaton Papers as well as selected materials scattered among the papers of DeWitt Clinton, George Clinton, James Hall, Henry Post, David Thomas, Elkanah Watson, and the Van Rensselaer Manor Papers.

The archives at Rensselaer Polytechnic Institute's library answered my search for the original of the final volume of Eaton's *Geological Journal*, and the Gerald Friedman manuscript collection there now contains an added bounty of materials pertaining to the history of geology in New York State. A visit to Yale's University archives in New Haven, Connecticut, gave me direct contact with Benjamin Silliman's manuscript collection, but I actually found that the two-volume *Life of Benjamin Silliman*, ed. George P. Fisher (New York: Charles Scribner and Company, 1866) provided a reasonably adequate window into those portions of his private papers and correspondence that I needed. Several fruitful visits to the American Antiquarian Society in Worcester, Massachusetts, yielded valuable auxiliary materials, such as the feisty letters and voluminous research manuscripts of Caleb Atwater.

A side trip to the Transylvania University archives in Lexington, which I made while attending a conference in Kentucky, enabled me to track down not only some of the primary source materials pertaining to Constantine Rafinesque's interactions with Amos Eaton and New York geology but also a little more information about Amos Eaton's son Hezekiah Hulbert Eaton. Transylvania had hired H. H. Eaton away from the Rensselaer School in 1829 to teach courses in chemistry and electricity to medical students and to carry forward his own researches in botany and paleontology in what was then the relatively sparsely settled western part of the United States. The younger Eaton worked for three years in Kentucky, only to succumb rather suddenly to the ravages of tuberculosis. The longest-lived of Eaton's three most scientifically talented offspring, Hezekiah died just three and half weeks after celebrating his twenty-third birthday.

Whereas scholars used to be obliged to undertake physical journeys to archives and repositories of historical documents, recent years have seen a rapid increase in the quality and quantity of digitally archived images and documents that are available through the Internet. Indications of source materials found by this means are given where appropriate in the endnotes.

Because many of my subjects enjoyed greater notoriety than did Amos Eaton, I have been able to mine a number of published works in search of useful primary sources that could help to round out this study. In particular, passages dealing with Washington Irving, William Cullen Bryant, DeWitt Clinton, and Joseph Henry draw heavily on the published papers of these leading figures in the history of American literature, politics, and science. These works typically present primary source materials in either complete or abridged formats that permit one to form one's own independent interpretations. These valuable compilations include Van Wyck Brooks, *The World of Washington Irving* (New York: E. P. Dutton, 1944); *The Letters of William Cullen Bryant*, ed. William Cullen Bryant II and Thomas G. Voss (New York: Fordham University Press, 1975); *The Life and Writings of DeWitt Clinton*, ed. William W. Campbell (New York: Baker and Scribner, 1849); and *The Papers of Joseph Henry*, ed. Nathan Reingold (Washington: Smithsonian Institution Press, 1972).

Fairly early in my thinking about this topic, I happened to stumble across three especially valuable primary sources. These books helped convince me that it made sense to range freely from a narrow history of science topic in order to document and contemplate the rich interconnections between the worlds of politics, the arts, and science in this period of American history. The catalysts for this decision were Timothy Dwight's collected letters, posthumously published in multiple volumes under the title *Travels in New-England and New-York* (New Haven: Timothy Dwight, 1822); DeWitt Clinton's pseudonymously authored book, Hibernicus, *Letters on the Natural History and Internal Resources of the State of New York* (New York: Bliss and White, 1822); and the account by the English scientific tourist John Finch, *Travels in the United States of America and Canada, containing some accounts of their scientific institutions, and a few notices of the geology and mineralogy of those countries, to which is added An Essay on the Natural Boundaries of Empires* (London: Longman, Rees, Orme, Brown, Green, and Longman, 1833). From that promising start, I began to gather travel accounts by scientific explorers, literary figures, and tourists of every stripe. (Many are cited among the chapter notes.) Once I recognized how integral geology could be to the lived experience of a certain class of intellectuals in the early American republic, I was emboldened to carry this project through to its present multidimensional state.

SECONDARY SOURCES

Thus broadly conceived, this book has had to sample a wide variety of historical scholarship. Within the history of American science, I have necessarily delved into narrative accounts that either featured Amos Eaton or early nineteenth-century New York geology, or both. The two most useful comprehensive works that addressed both of these criteria are Ethel M. McAllister, "Amos Eaton, Scientist and Educator" (Ph.D. diss., University of Pennsylvania, 1941); and the book based on Michele L. Aldrich's brilliant doctoral research, *New York State Natural History Survey, 1836–1842* (Ithaca, NY: Paleontological Research Institution, 2000). I have also found it helpful to resort to various other specialized articles, such as Harlan Hoge Ballard, "Amos Eaton: A Pioneer of Science in Berkshire County," *Collections of the Berkshire Historical and Scientific Society* (Pittsfield, MA: Sun Printing Co., 1897): 179–312; H. S. Van Klooster, "Amos Eaton as a Chemist," *Journal of Chemical Education* 15 (1938): 453–60; and Samuel Reznek, "A Traveling School of Science on the Erie Canal in 1826," *New York History* 40 (1959): 255–69.

Among classic scholarly works touching on various aspects of the history of American natural history, geology, or the nature of New York State, I consider the following to be mostly reliable (if sometimes provocative) contributions: John M. Clarke, *James Hall of Albany: Geologist and Paleontologist (1811–1898)* (Albany, 1923); Dirk J. Struik, *Yankee*

Science in the Making: Science and Engineering in New England from Colonial Times to the Civil War (New York: Dover Books, 1948/1991); Jeannette E. Graustein, *Thomas Nuttall Naturalist: Explorations in America 1808–1841* (Cambridge, MA: Harvard University Press, 1967); John C. Greene, *American Science in the Age of Jefferson* (Ames: Iowa State University Press, 1984); Bradford B. Van Diver, *Roadside Geology of New York* (Missoula, MT: Mountain Press, 1985); William Wyckoff, *The Developer's Frontier: The Making of the Western New York Landscape* (New Haven: Yale University Press, 1988); Edmund Berkeley and Dorothy Smith Berkeley, *George William Featherstonhaugh: The First U.S. Government Geologist* (Tuscaloosa: University of Alabama Press, 1988); Simon Baatz, *Knowledge, Culture, and Science in the Metropolis: The New York Academy of Sciences, 1817–1970* (New York: The New York Academy of Sciences, 1990); Chandos Michael Brown, *Benjamin Silliman: A Life in the Young Republic* by (Princeton: Princeton University Press, 1998).

Valuable scholarship on these general topics continues to be produced. Among more recently published works, the following contrast with and complement my own work: Conevery Bolton Valenčius, *The Health of the Country: How American Settlers Understood Themselves and Their Land* (New York: Basic Books, 2002); Paul Lucier, *Scientists and Swindlers: Consulting on Coal and Oil in America, 1820–1890* (Baltimore: Johns Hopkins University Press, 2008); Sara S. Gronim, *Everyday Nature: Knowledge of the Natural World in Colonial New York* (New Brunswick, NJ: Rutgers University Press, 2009); Benjamin R. Cohen, *Notes From the Ground: Science Soil and Society in the American Countryside* (New Haven: Yale University Press, 2009); Richard W. Judd, *The Untilled Garden: Natural History and the Spirit of Conservation in America, 1740–1840* (Cambridge: Cambridge University Press, 2009); Sally Gregory Kohlstedt, *Teaching Children Science: Hands-On Nature Study in North America 1890–1930* (Chicago: University of Chicago Press, 2010); David Stradling, *Making Mountains: New York City and the Catskills* (Seattle: University of Washington Press, 2010) and *The Nature of New York: An Environmental History of the Empire State* (Ithaca, NY: Cornell University Press, 2010); and Andrew J. Lewis, *A Democracy of Facts: Natural History in the Early Republic* (Philadelphia: University of Pennsylvania Press, 2011), which offers an interpretation of early American natural history and its social implications that focuses more on social status and the shifting dynamics of natural history's elitism than the argument I have presented in this book about geology's intrinsic interest for citizens of early nineteenth-century New York.

For background information on those early political leaders who took a special interest in the natural sciences in general (and geology in particular), I consulted the following mostly biographical works: Samuel L. Mitchill, *A Discourse on the Character and Services of Thomas Jefferson, more especially as a promoter of natural and physical science. Pronounced, by request, before the New York Lyceum of Natural History, on the 11th October, 1826* (New York: G. and C. Carvill, 1826); Samuel L. Mitchill, *Discourse on the Character and Scientific Attainments of DeWitt Clinton, late Governor of the State of New-York; pronounced at the Lyceum of Natural History, of which he was an honorary member, on the 14 July, 1828* (New York: E. Conrad, 1828); David Hosack, *Memoir of DeWitt Clinton, with Illustrations of the Principal Events of His Life*, 2 vols. (New York,

1829); James Renwick, *Life of Dewitt Clinton* (New York: Harper and Brothers, 1840); Dorothie Bobbe, *DeWitt Clinton* (New York: Minton, Balch, and Co., 1933); Daniel J. Boorstin, *The Lost World of Thomas Jefferson* (Boston: Beacon, 1948; repr., Chicago: University of Chicago Press, 1993); Edward T. Martin, *Thomas Jefferson, Scientist* (Henry Schuman Inc., 1952; repr., New York: Collier Books, 1962); Jonathan Daniels, *Ordeal of Ambition: Jefferson, Hamilton, Burr* (New York: Doubleday, 1970); Charles A. Miller, *Jefferson and Nature: An Interpretation* (Baltimore: Johns Hopkins University Press, 1988); I. Bernard Cohen, *Science and the Founding Fathers: Science in the Political Thought of Thomas Jefferson, Benjamin Franklin, John Adams and James Madison* (New York: W. W. Norton, 1995); Evan Cornog, *The Birth of Empire: DeWitt Clinton and the American Experience, 1769–1828* (New York: Oxford University Press, 1998); James J. Kirschke, *Gouverneur Morris: Author, Statesman, and Man of the World* (New York: St. Martin's Press, 2005); and Edward J. Larson, *A Magnificent Catastrophe: The Tumultuous Election of 1800, America's First Presidential Campaign* (New York: Free Press, 2007).

The construction of the Erie Canal and the era of internal improvements in the United States provide a centerpiece to this study of how science, technology, and society all came together as mutually reinforcing elements of New York State's rise to power, prominence, and prosperity. The literature on the history of the Erie Canal is simply massive. Though hardly any of these works pay attention to the geological sciences, I still found them to be highly useful sources of historical context and insights into the economic, political, and social consequences of the canal: Noble E. Whitford, *History of the Canal System of the State of New York: Together with brief histories of the canals of the United States and Canada* (1905; repr., Albany: Brandow Publishing, 1906); Mary Kay Phelan, *Waterway West: The Story of the Erie Canal* (New York: Thomas Y. Crowell Co., 1977); Ronald E. Shaw, *Canals For a Nation: The Canal Era in the United States, 1790–1860* (Lexington: University Press of Kentucky, 1990); Carol Sheriff, *The Artificial River: The Erie Canal and the Paradox of Progress* (New York: Hill and Wang, 1996); John Lauritz Larson, *Internal Improvement: National Public Works and the Promise of Popular Government in the Early United States* (Chapel Hill: University of North Carolina Press, 2001); and Andro Linklater, *Measuring America: How an Untamed Wilderness Shaped the United States and Fulfilled the Promise of Democracy* (New York: Walker and Co., 2002).

The chapters analyzing the interest of early nineteenth-century literary figures look to significant works of scholarship in that area of specialized historical study, in addition to the primary sources mentioned earlier. Notably relevant sources on early New York writers include Donald A. Ringe, "William Cullen Bryant and the Science of Geology," *American Literature* 26 (1955): 507–14; Norman Foerster, *Nature in American Literature: Studies in the Modern View of Nature* (New York: Russell and Russell, 1958); Harry Hayden Clark, "Fenimore Cooper and Science," *Transactions of the Wisconsin Academy of Sciences, Arts and Letters* 48 (1959): 179–204 and 49 (1960) 249–82; William L. Hedges, *Washington Irving: An American Study, 1802–1832* (Baltimore: Johns Hopkins University Press, 1965); Lewis Leary, *Soundings: Some Early American Writers* (Athens: University of Georgia Press, 1975); Alan Taylor, *William Cooper's Town: Power and Persuasion on the Frontier of the Early American Republic* (New York: Vintage Books, 1995); Andrew

Burstein, *The Original Knickerbocker: The Life of Washington Irving* (New York: Basic Books, 2007); Wayne Franklin, *James Fenimore Cooper: The Early Years* (New Haven: Yale University Press, 2007); and Gilbert H. Muller, *William Cullen Bryant: Author of America* (Albany: SUNY Press, 2008).

Works by scholars that support analysis of the relationship between nineteenth century visual arts and contemporary geological thinking include Barbara Novak, *American Painting of the Nineteenth Century: Realism, Idealism, and the American Experience* (New York: Harper and Row, 1979), and *Nature and Culture: American Landscape and Painting, 1825–1875* (New York: Oxford University Press, 1980); Ellwood C. Parry III, *The Art of Thomas Cole: Ambition and Imagination* (Newark: University of Delaware Press, 1988); Stephen Jay Gould, "Church, Humboldt, and Darwin: The Tension and Harmony of Art and Science," in *Frederic Edwin Church*, ed. Franklin Kelly (Washington: Smithsonian Institution Press, 1989); *Thomas Cole: Drawn to Nature* (Albany: Albany Institute of History and Art, 1993); Kenneth Haltman, "The Poetics of Geologic Reverie: Figures of Source and Origin in Samuel Seymour's Landscapes of the Rocky Mountains," *Huntington Library Quarterly* 59 (1996): 303–47; Rebecca Bedell, *The Anatomy of Nature: Geology and American Landscape Painting, 1825–1875* (Princeton: Princeton University Press, 2002); Andrew Graciano, " 'The Book of Nature is Open to All Men': Geology, Mining, and History in Joseph Wright's Derbyshire Landscapes," *Huntington Library Quarterly* 68 (2005): 583–600; and Linda S. Ferber, *The Hudson River School: Nature and the American Vision* (New York: New-York Historical Society, 2009).

Lastly, the controversial subject of religion and science has generated a gigantic literature of its own. I can point to some helpful writers whose examinations of nineteenth-century geology extend the discussion that I have barely broached in the concluding section of this book. The classic work, Charles Coulston Gillispie, *Genesis and Geology: A Study in the Relations of Scientific Thought, Natural Theology and Social Opinion in Great Britain, 1790–1850* (1951; repr., Cambridge: Harvard University Press, 1996) is a good place to start. Other excellent studies include Rodney Lee Stiling, "The Diminishing Deluge: Noah's Flood in Nineteenth-century American Thought" (Ph.D. diss., University of Wisconsin-Madison, 1991); Ronald L. Numbers, *The Creationists: The Evolution of Scientific Creationism* (Berkeley: University of California Press, 1993); and Walter H. Conser Jr., *God and the Natural World: Religion and Science in Antebellum America* (Columbia: University of South Carolina Press, 1993). Key works that focus on the upsurge of religious revivalism in the Erie Canal district include Whitney R. Cross, *The Burned-Over District: The Social and Intellectual History of Enthusiastic Religion in Western New York, 1800–1850* (Ithaca: Cornell University Press, 1950); Paul E. Johnson, *A Shopkeeper's Millennium: Society and Revivals in Rochester, New York, 1815–1837* (New York: Hill and Wang, 1979); Michael Barkun, *Crucible of the Millennium: Burned-Over District of New York in the 1840s* (Syracuse: Syracuse University Press, 1986); and John L. Brooke, *The Refiner's Fire: The Making of Mormon Cosmology, 1644–1844* (Cambridge: Cambridge University Press, 1996).

Index

Decatur, Stephen, 157
Deism, 29, 223n17
De Kay, James, 177, 179
Deluge, the: as geological agent, 31, 33,
   96, 109, 111–114, 119–121, 197, 199–201,
   203; as subject of art, 187–191, 249n16
democratic principles, 13, 33, 37, 40, 84, 94
Democrat-Republican Party, 2, 38, 40,
   155, 179
Dewey, Chester, 70–71
De Witt, Simeon, 77, 86
differences between Old and New
   Worlds, 42, 54, 75, 111, 133, 139, 148, 173
diluvialism, 20, 109, 236n28; in Cole's
   art, 188–190, 195; as Eaton's systematic
   theory, 110–111, 114, 118–121, 124–126, 138,
   149, 182, 215; as naïve way to interpret
   glacial evidence, 95–96, 238n16; and
   religion, 198–201, 207, 209, 211
Ditton, John, 210
drift, glacial. See erratic boulders
dueling, 36–37, 46–47, 157
Dunlop, William, 179, 186
Durand, Asher B., 179, 186–187, 195
Dwight, Timothy: as geological observer,
   29–32, 34, 101; as Yale president, 26, 45,
   68, 177, 197, 214–215, 223nn17–18

earthquakes, 20, 28, 57, 182
Eaton, Amos: as botanist, 5, 68–72, 74,
   168; early years of, 5–6, 66–80, 83,
   154–155, 166, 177, 198, 241n3, 251n6; as
   educator, 6, 8, 70–79, 115–116, 122–124,
   126, 164–165, 168–173, 196, 199, 203,
   205–206, 229n28, 243n35; as geologist/
   mineralogist, 6–7, 9, 11–12, 25, 35, 63,
   69–70, 72–80, 94–127, 132–149, 169, 175,
   182, 188–190, 193–195, 198–210, 219n7,
   220n11, 229n29, 234n5, 236n38, 237n43,
   238n16, 239n43; historiography of, 7–8,
   73, 210–211, 220n14, 228n13, 230n41,
   237n44, 240n50, 241n61; influence of,
   on American science, 5–7, 26, 72–74,
   101, 142–143, 149, 210–216
Eaton, Anna Bradley, 136
Eaton, Charles Linnaeus, 68
Eaton, Hezekiah Hulbert, 124, 126
Eaton, Polly Thomas, 66–67, 69

Eaton, Sarah (daughter of Amos), 171
Eaton, Sarah (Sally) Cady, 66–68, 70
Eaton, Timothy Dwight, 68, 124
Eaton, William, 171
Eddy, Thomas, 86–87, 90
Edwards, Jonathan, 45, 198
Eights, James, 124
elections: of 1800, 41, 43, 47, 227n42; of
   1812, 2, 48; of 1816, 3, 76; of 1817 (New
   York), 3, 76, 89–90, 219n3, 230n32; of
   1824, 60, 64, 127, 137, 227n42
Ellicott, Andrew, 27, 89–90
Ellicott, Joseph, 88–89
Ellison, Thomas, 176
Ellsworth, Henry Leavitt, 159–161
Embargo Act (1807), 86, 166, 178
Emmons, Ebenezer, 6, 72, 126, 149, 254n5
Empiricism, 33, 61, 102, 110, 189, 203, 206,
   208; as preferred scientific method, 26,
   75, 79, 200
Enlightenment, the, 3, 13, 33, 42, 58, 153,
   165, 208, 217, 232n16; in France, 84; in
   Scotland, 43, 204
Erie Canal: historiography of, 8, 36, 115;
   planned construction of, 2, 83–84,
   87–88, 90–92, 237n45; as scientific
   opportunity, 3, 6, 83, 94–95, 97–98,
   100, 103–104, 106–107, 110, 124, 133, 135,
   143, 145, 148, 210, 254n5; success of, 12,
   117–119, 127–129, 131, 198, 212
erratic boulders, 32, 97, 101, 182, 193, 195,
   201, 203, 207
evolution, 203, 211

facts in science, 35, 57, 99; contested
   nature of, 6, 11, 52, 95, 203, 205,
   214–215, 229n29
Falls of the Ohio, 85
Fanning, Edmund, 124
Faraday, Michael, 3
Featherstonhaugh, George William,
   136–140, 147–148, 187, 214, 240n50,
   248n8
Federalist Party, 2, 38–41, 43, 45–48, 64,
   69, 84, 198, 204, 230n32; and early
   writers, 155–156, 166, 175–176; and Erie
   Canal, 86–87, 90; and Rensselaer, 60,
   64, 212